Effective Building Maintenance:
Protection of Capital Assets

Effective Building Maintenance:
Protection of Capital Assets

Herbert W. Stanford III, PE

Routledge
Taylor & Francis Group
LONDON AND NEW YORK

Published 2020 by River Publishers

River Publishers

Alsbjergvej 10, 9260 Gistrup, Denmark

www.riverpublishers.com

Distributed exclusively by Routledge

4 Park Square, Milton Park, Abingdon, Oxon OX14 4RN

605 Third Avenue, New York, NY 10017, USA

Library of Congress Cataloging-in-Publication Data

Stanford, Herbert W.

Effective building maintenance : protection of capital assets / Herbert W. Stanford III

p. cm.

Includes bibliographical references and index.

ISBN-10: 0-88173-638-4 (alk. paper)

ISBN-13: 978-8-7702-2286-0 (electronic)

ISBN-13: 978-1-4398-4553-0 (Taylor & Francis : alk. paper)

1. Buildings--Maintenance. I. Title.

TH3351.S73 2010

658.2′02--dc22

2010019108

Effective building maintenance : protection of capital assets / Herbert W. Stanford III

First published by Fairmont Press in 2010.

Routledge is an imprint of the Taylor & Francis Group, an informa business

10: 0-88173-638-4 (The Fairmont Press, Inc.)

13: 978-1-4398-4553-0 (print)

13: 978-8-7702-2286-0 (online)

13: 978-1-0031-5155-5 (ebook master)

While every effort is made to provide dependable information, the publisher, authors, and editors cannot be held responsible for any errors or omissions.

Table of Contents

Chapter 1

Facilities As Assets

"...You cannot escape the responsibility of tomorrow
by evading it today..."

Abraham Lincoln

This text attempts to address, in as much detail as possible, the requirements for designing, implementing, and managing programs and procedures for the maintenance of major building elements from the foundation to the roof, including interior and exterior support systems and sitework elements. However, there are three "maintenance" aspects that this text does not include since there are already numerous texts available that address these topics in detail:

Custodial services: Obviously, routine indoor housekeeping is needed, but it should not be considered part of the facility maintenance program—just because the facility is clean, that does not mean it is not suffering from lack of maintenance needed to protect the facility's asset value.

Landscape services: Landscape services, grass cutting, etc., is routine outdoor housekeeping that, also, should not be considered part of the facility maintenance program. Appendix G does include specific maintenance procedures for most facility site features—berms, swales, retention ponds, etc.—and Chapter 4 offers specific recommendations relative to outdoor water consumption and site stormwater runoff, which can be addressed by improved landscaping concepts.

Safety and health issues: Maintenance procedures must incorporate appropriate safety and health considerations including everything from ladders and lifting to confined spaces. The details of these requirements and procedures to be responsive to them are beyond the scope of this text. However, every maintenance program must include the required procedures for effective lockout/tagout (LOTO). For mechanical devices, OSHA Standard 29 CFR 1910.47 must be met, while, for electrical maintenance, the procedures defined in NFPA Standard 70E are mandatory.

PROGRAMMED MAINTENANCE AS ASSET MANAGEMENT

Critical infrastructure and industrial facility owners and operators have adopted the term "asset management" to describe their core role in life—caring for and obtaining a satisfactory level of service from the physical plant, infrastructure, and associated facilities. Their concept is that, *since facilities represent significant capital assets, they must be protected through well-planned and appropriately funded programmed maintenance.*

For industrial and many commercial facilities (retail, offices, etc.), the level of maintenance can be evaluated as part of an overall optimization of asset utilization, balancing the cost of maintenance and retained value of the asset against the overall economics of the business "model." However, for public and institutional facilities, the facility itself is often the most important asset and the need for maintenance at a level that allows the facility design life to be attained and its economic value be retained is paramount. In the public and institutional environment, "cash flow" rarely supports significant renovation or replacement of facilities as it does in the industrial and commercial sectors. The starting point for implementing an asset-based maintenance program is to understand two important concepts.

First, as shown in Figure 1-1, the goal of maintenance is to prevent, or at least reduce, the degradation or deterioration of the quality of service provided by each building component over its design service life. Maintenance is not a process of improvement that may be required to meet increased expectations for performance beyond which the component was originally designed to provide.

Second, each building, and the individual components that make-up that building, has a finite service life. At the end of that life, replacement or major renovation is required since, in the long run, it is less expensive to replace or renovate than to continue to make repairs that become increasingly frequent and costly.

Programmed maintenance, as illustrated by Figure 1-2, is a combination of preventative maintenance and planned replacement or major renovation of building components.

Planned replacement or major renovation is the step taken when a component reaches the end of its design service life and is discussed in Chapter 2 of this text.

Preventative maintenance ensures that facility components actually achieve their service life and consists of two elements:

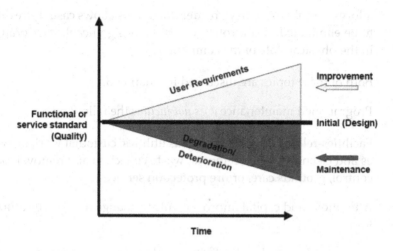

Figure 1-1. Functionality/Quality vs. Time

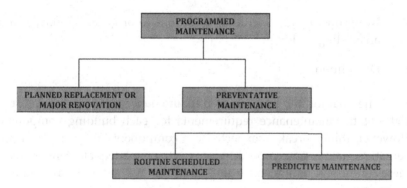

Figure 1-2. Elements of Programmed Maintenance

1. Routine maintenance, which consists of specific procedures that are performed on a regular schedule. These procedures are designed to detect, preclude, or mitigate degradation of a facility system (or its components). The goal of routine, scheduled maintenance is to minimize each component's degradation and thus maintain, or even extend, the useful life of the component.

2. Predictive maintenance uses routine inspection and evaluation, testing, and analysis to augment routine, scheduled maintenance procedures by detecting the onset of component degradation and to

address problems as they are identified. This allows casual stressors to be eliminated or controlled prior to any significant deterioration in the physical state of the component.

Both of these topics are discussed in Chapter 3.

Programmed maintenance *does not include* the following:

1. Facilities-related operations such as utilities, custodial work (services and cleaning), snow removal, waste collection and removal, pest control, grounds care, or fire protection services.

2. Alterations and capital improvements to change function or utilization.

3. Legislatively mandated activities such as improvements for accessibility, dealing with hazardous materials, etc.

4. New construction, including additions to or general renovation of an existing building.

5. Demolition.

The goal for any programmed maintenance program is to address 100% of the maintenance requirements for each building component. However, things break unexpectedly, a component fails early, or a preventative maintenance procedure proves to be inadequate. Under any of these conditions, maintenance and repair that has not been programmed is required.

Unprogrammed maintenance can be identified by a building occupant/user, by the maintenance staff while performing routine preventative maintenance procedures, or by predictive maintenance tests and evaluations. The result is that a work order must be issued for the necessary unprogrammed, but needed, maintenance or repair. (At its worse, unplanned component replacement or major renovation may even be required.)

Thus, it is essentially impossible to achieve the goal that 100% of maintenance activities be programmed—a 70-80% goal is more realistic. *If unprogrammed maintenance consumes 30% or more of the available maintenance resources, then there are simply inadequate resources available to provide the required level of programmed maintenance.*

MAINTENANCE PLANNING AND BUDGETING

In establishing (or re-establishing) a maintenance department, specific maintenance factors for each facility must be evaluated. These factors include level of maintenance to be provided, how does facility use affect maintenance requirements, and should maintenance be provided by outside contractors or "in-house" staff. Once these decisions are made, a realistic budget for maintenance can be established.

Maintenance Staffing Guidelines for Educational Facilities, published in 2002 by the Association of Physical Plant Administrators or APPA (now the Association of Higher Education Facilities Officers) defines levels of maintenance in accordance with the following:

Table 1-1

Maintenance Level	Description
1	Showpiece Facility
2	Comprehensive Stewardship
3	Managed Care
4	Reactive Maintenance
5	Crisis Response

For most facilities, the minimum acceptable maintenance level corresponds to APPA's Level 2 ("Comprehensive Stewardship"). APPA defines this level of maintenance as follows:

"Maintenance activities appear organized with direction. Equipment and building components are usually functional and in operating condition. Service and maintenance calls are responded to in a timely manner. Buildings and equipment are regularly upgraded, keeping them current with modern standards and usage."

Obviously, this level of maintenance recognizes that each facility is an asset that must be maintained in order to achieve maximum return on investment. **This text, then, is based on providing APPA's Level 2 maintenance for all facilities.**

For maintenance staffing and budgeting purposes, facilities can be divided into categories based on their "complexity":

Type 1 facilities include K-12 schools, office buildings, retail buildings, hotels, nursing homes, medical clinics, etc. that are generally relatively basic buildings designed for a relatively short to medium life (30 years or less). Stand-alone buildings in this category generally will make use of contracted maintenance services. However, if considered in aggregate—a school system, a chain of stores, a multi-building office campus—the need for an "in-house" maintenance staff will normally arise, even though some maintenance needs may still be met on a contract basis.

Type 2 facilities include institutional facilities (such as higher education classroom and administrative buildings, government offices, etc.) and basic industrial buildings. These buildings are designed for longer life (40-50 years), have more complex systems and components, and represent higher asset value to the user.

Type 3 facilities include hospitals, research laboratories, data centers, and high technology manufacturing. These buildings have very complex systems and components and require much higher levels of reliability than either Type 1 or Type 2 facilities.

Generally, only an in-house maintenance program will provide the level of programmed maintenance typically required for larger, and especially more complex, facilities. However, even for larger facilities, special area maintenance may be provided by outside contractors for elevators, fire sprinkler systems, laboratory fume hoods, security systems, HVAC controls, etc. simply because only the very largest facilities can afford maintain full-time specialist maintenance technicians to meet the maintenance requirements of these components. This approach can be incorporated into any maintenance department organization and staffing plan.

To achieve APPA Level 2 maintenance, appropriate budgeting and funding for maintenance must be established. There are two approaches to this:

1. The National Research Council (NRC) studied the maintenance of Federal facilities in the late 1990's (the results of which were published in 1998 as *Stewardship of Federal Facilities*). This study found that for an acceptable level of maintenance, the preventative maintenance budget was 2-4% of the facility's current replacement cost. This amount does not include the costs of performing maintenance deferred from prior years.

    For smaller facilities with limited planning and budgeting resources, use of the NRC approach for maintenance budgeting may be an good initial approach and the recommended level of maintenance can be expected to require annual funding of 3.5-4% of current replacement cost.

2.   A better approach is to use "zero-based budgeting" whereby the actual annual costs of each required maintenance element are determined. This approach has the benefit of defining not only costs, but it provides excellent guidance as to the staffing needs for implementing these procedures.

There are numerous annual surveys of maintenance expenditures by facilities published by Whitestone, Building Owners and Managers Association (BOMA), APPA, etc. that define "what the other guys are doing." While interesting, these data are for all practical purposes worthless since there are no corresponding data as to the effectiveness of these expenditures. (However, they may provide some basis for justifying maintenance budgets to upper management!)

STAFFING THE MAINTENANCE DEPARTMENT

With the exception of the APPA publication discussed above (and some individual state recommendations for K-12 schools), there are no general maintenance staffing standards in place in the United States. To define staffing needs, each facility, the components of that facility, and required component maintenance procedures must be carefully evaluated.

Any maintenance department (and note that we are excluding custodial and grounds services from this discussion) can be organized generally along the staffing levels shown in Table 1-2.

Basic job descriptions for each of these staff positions are provided at the end of this chapter.

The number of each type of staff member required at any facility will vary based on the size and complexity of the facility and the level of contracted maintenance services utilized by the facility. The relationship between the size (gross area) of a facility and the types of buildings and other elements (complexity) making up the facility will determine the final mix of types and number of staff positions. But, in general, the following guidelines can be applied:

Table 1-2

Maintenance Management	Facilities Maintenance Manager Assistant Facilities Maintenance Manager
Craft Supervisor	General Facility Maintenance Supervisor HVAC Maintenance Supervisor Plumbing / Fire Protection Maintenance Supervisor Electrical Maintenance Supervisor
Craft Technicians (General Facility, HVAC, Plumbing, Electrical)	Technician Level I Technician Level II Technician Level III

Special
Technicians
(All Level III)
	General Facility: Carpenter / Cabinetry
	    Painter
	    Locksmith
	    Conveying Systems Technician
	    Masonry / Tile
	    Roofer
	
	HVAC: Controls Technician
	Fire Protection: Fire Sprinkler Technician
	Electrical: Special Systems Technician (Fire
	    Alarm, Security, and Communications)

Clerical	Administrative Assistant General Clerical Work Order System Specialist

1. Every maintenance department must have a manager. Since the maximum number of people that can be effectively managed by one individual is 10, assistant maintenance managers and/or craft supervisors must be added as the size of the department increases.

2. Assistant maintenance managers are required when the department must be organized on a functional basis. For example, on assistant manager may be responsible for maintenance planning, another for maintenance procedures, a third for staff development and training, etc. Typically, only the larger facilities require assistant managers.

3. A craft supervisor is typically required when the number of technicians in a given craft area exceeds 5-6. (Once there are more than 3 technicians in a craft area, one Level III technician may be appointed "foreman.")

4. The following table reflects the staffing level for craft technicians that is typically required per 100,000 gsf of each type of facility. However, these levels may be increased or decreased due to many factors, including staff qualifications and productivity, the quality of the facility construction, previous deferred or avoided maintenance, etc.

Table 1-3. Typical Maintenance Staffing Levels

Technical Staff	Type 1 Facility	Type 2 Facility	Type 3 Facility
General Facility Technician	0.4	0.5	0.5
HVAC TEchnician	0.8	1.1	1.5
Plumbing Technician	0.1	0.2	0.2
Electrical Technician	0.3	0.4	0.5
Total	**1.6**	**2.2**	**2.7**

Staffing in each craft area should typically be allocated with 10-20% as Level I technicians, 20-30% as Level II technicians, and 50-70% as Level III technicians. This provides for staff development and growth and transfer of "institutional memory" and makes hiring easier as higher level technician turnover occurs.

5. Special technicians may be required based on size of the facility, the special elements of the facility, and the ability (or not) of obtaining required specialized maintenance services under contract locally at an affordable price.

6. Administrative support requirements will vary with need. One element that is absolutely critical is to have at least two employees in this area—one with general clerical responsibilities and one with

responsibility for managing the work order or maintenance management system. Two people means the phones in the maintenance department will be "manned" for at least 10-12 hours/day by using staggered work and break schedules.

Too often, the maintenance department becomes the "construction department," assigned responsibility by management for replacement/renovation projects or even construction of new facilities. When this happens, programmed maintenance always suffers!

There is no easy solution to this problem once a maintenance department begins to perform construction work. The best approach is to head-off the problem before it becomes one by making it clear to management that current staffing levels are predicated on maintenance needs and assigning staff to renovation or new construction projects is very much a game of "robbing Peter to pay Paul." There are two potential negative results from this problem:

1. If current staffing levels are maintained, some programmed maintenance will be deferred.

2. If additional maintenance staff is hired to offset personnel losses to renovation or new construction projects, when these projects are completed, the maintenance department is then over-staffed, adding unnecessary cost to the department's operating budget.

The best solution is to keep maintenance people performing maintenance procedures and to hire outside contractors to perform renovation or new construction projects.

TARGETED MAINTENANCE TRAINING

The development and implementation of a maintenance skills training program must be part of a well-developed maintenance management program. Training for the sake of training is a waste since skill increases that are not utilized properly will result in no changes. Once an individual is trained in a skill, he must be provided with the time and tools to perform this skill and must be held accountable for his actions.

To implement an effective training program, the first step to perform a "skills assessment" to identify current skill deficiencies. From numerous studies, it is well known that the skills levels of most maintenance personnel fall short of what is acceptable. One professional training firm found that 80% of maintenance personnel had less than 50% of the basic technical skills to perform their jobs! [Study by the Technical Training Division, Life Cycle Engineering, North Charleston, South Carolina]. Thus, it is a given that an effective training program is *always needed* as part of a stewardship maintenance program.

A detailed assessment of current maintenance skills will dictate a "targeted," non-redundant training program that addresses skill improvements needed in order for the technician to perform his or her job effectively. The skill assessment should have three components:

1. Written test that identifies the knowledge required for a specific skill. The test must Incorporate questions about theories, principles, fundamentals, vocabulary, and calculations required as part of the skill set. For example, an HVAC technician who can't define the difference between "relative humidity" and "humidity ratio" won't be able to diagnose and correct humidity control problems in a building. Likewise, a brick mason who can't determine the difference between a step flashing and a pan flashing, is not a good brick mason.

2. Identification test that assesses "hands-on" knowledge in specific skill areas. Employees, during a tour of the facility, are asked to name components and explain their uses in an oral assessment.

3. Performance evaluation. To analyze this aspect, employees carry out typical maintenance procedures while be observed by supervisors or consultants.

The assessment results, both on an individual basis and on a maintenance department basis, can then be used for two functions:

1. Define the needed training to improve the skill levels of each individual and to ensure the department has the requisite skill resources available to support a stewardship maintenance program.

2. Provide the basis of "performance standards" to used in job descriptions for new hires and to evaluate individual employee per-

formance and, most importantly, his or her improvement resulting from a training program.

Training costs money. There is the cost of the training itself, but there is a much greater cost associated with the "downtime" expended during training sessions. This will cost will, inevitably, attract the attention of management who will ask the following questions:

1. How much will the training cost?

2. How much money will be saved by implementing the training program?

The answer to the first question will depend on the results of the needs identified by the skills assessments. But, the answer to the second question can be found in the study funded by the U.S. Department of Education and the U.S. Census Bureau to determine how training impacts productivity:

1. Increasing an individual's educational level by 10% increases that individual's productivity by 8.6%.

2. Increasing an individual's work hours by 10% increases that individual's productivity by 6.0%.

3. Increasing capital investment by 10% increases productivity by 3.2%.

Thus, training is the most cost effective method of improving productivity with any employee!

OUTSOURCING: CONTRACT MAINTENANCE

The use of outsourcing or contract maintenance is fairly common in industrial plants, primarily due to the emphasis on production equipment reliability and the need to minimize production downtime. Likewise, contract maintenance is used to a fairly significant extent in commercial facilities. A survey by *FMLink*, an online facilities publication, in

2002 indicated that, while contracting for unskilled/semi-skilled services such as housekeeping and landscape services was very high (65-75% of all businesses), the use of outsourcing for preventative maintenance was much lower (50% or less). For public owners, hospitals, public schools, colleges and universities, and other institutions the use of contracted preventative maintenance was below 20%.

So, when do you make the decision to switch from in-house maintenance to contract maintenance?

As discussed previously, unprogrammed maintenance produces the need to immediately respond to maintenance problems. This, in turn, creates wide fluctuations in the work load imposed on the maintenance staff. So, the better job done reducing unprogrammed maintenance through a good preventative maintenance program and well-established maintenance priorities, the less variability there will be in staff workload from week to week.

Large variations in maintenance workload due to unprogrammed maintenance will lead to poor utilization of resources and, potentially, overstaffing. When this happens, someone starts talking about outsourcing. But, outsourcing maintenance resources under these conditions will change nothing. The maintenance contractor, to be effective, must provide better maintenance management to reduce unprogrammed maintenance. Otherwise, they will not be able to perform any better than the existing staff. *If this is the case, you must ask yourself why you cannot improve things if your contractor can?*

One answer may be that you have tried to make changes in the past, but the changes were vetoed by management or, if implemented, died a natural death due to staff resistance, organizational gridlock, or internal politics. In this case, outsourcing may be an "act of desperation" used to circumvent entrenched impediments to internal change and improvement. But, that is no guarantee of success; too often the scope of maintenance is limited by budgets and the contractor cost to implement a truly effective maintenance program may be far more than management thinks it can afford!

A better answer is that there is no reason you cannot implement an effective management plan! And, this better plan may include specific elements of outsourced maintenance for the following:

1.   **Special Skills**: Certain areas of maintenance require special skills, tools, equipment, etc. that can be expensive to acquire and even

more expensive to maintain. And, too often, even if in-house staffers have the needed special skills, these skills are used so infrequently that they cannot be maintained at a high level.

In these cases, outsourcing is more efficient and less costly. Maintenance of elevators and other conveying system, fire protection systems (sprinklers, fire suppression systems, and fire extinguishers), fire alarm and security systems, locks, and major HVAC components such as water chillers, cooling towers, and boilers are routinely outsourced. Direct digital controls systems are another component for which vendor maintenance is typically needed.

2. **Shutdowns and Outages**: It makes both common sense and eco-nomic sense to use contract maintenance services for scheduled shutdowns and outages. While more applicable for industrial equipment maintenance, it does apply to annual and semi-annual maintenance for major HVAC components, electrical switchgear, etc.

Vendors and contract maintenance companies often argue that they can provide preventative maintenance at a lower cost. Assuming that the in-house maintenance department is not over-staffed, for this to be true the contractor must use less skilled (lower wage rate) staff to perform maintenance functions, perform preventative maintenance at a reduced level from that currently provided (or that should be provided), or has reduced overhead costs (which may, in fact, be a reasonable assumption). Where an existing maintenance department has become bloated and/or inefficient due to excessive unprogrammed maintenance and/or its being used as a construction department, the cost savings produced by outsourcing can be easily matched by improved internal maintenance management.

Another vendor argument is that maintenance is not part of the "core" business of the organization and, thus, cannot be managed prop-erly within the business and/or it consumes resources (in addition to cash) that the business could better apply to its core activities. But, this is another bogus argument...it the building assets are not maintained, this negatively impacts the entire business, be it commerce, education, government, or health.

Outsourcing of preventative maintenance has serious potential risks that must be considered:

1. How will you deal with the divided loyalties of the contractor's staff? While they have an obligation to meet the contract requirements, the need to reduce costs imposed on their employer may result in maintenance objectives not being met.

2. How will the contractor protect you from being responsible for workplace safety and liability? The contractor's staff will be working in your facilities and you may be held accountable for workplace safety.

3. How will the requirements of applicable standards for indoor air quality, noise, energy, etc. be met and that compliance be documented? Without that being clearing defined, the facility owner may be responsible for health problems, etc. created by the contractor's failure to comply with applicable standards.

Thus, while contract maintenance may have a special place in any maintenance program, it will rarely be the total maintenance answer for most owners.

Once the need for some aspect of contract maintenance is identified, the contract for outsourced maintenance must define the following:

1. The level of maintenance required, specific maintenance procedures, and the method(s) of evaluating the service(s) provided by the contractor.

2. The contractor will be held responsible for not only the value of maintenance not done, or done poorly, but will be held responsible for the cost of the resulting degradation of building components and their shortened service life.

All maintenance contracts must be based on a fixed price, not man-hour rates. If the contract is based on technician rates, there is no incentive for the contractor to perform better since the more hours they sell, the more money that make.

A maintenance contract should be long term, not less than five years. Two reasons for this are incorporated in what Dr. Deming called the "seven deadly diseases" common to U.S. management: (1) "Lack of constancy of purpose" and (2) "Mobility of top management." Typically, the first phenomenon leads to the other. New managers are called in for fast and, unfortunately, often temporary results. They often change the organization, perhaps only because they want to bring in their buddies, make some cut backs, and then move on to another place before the long-term effects are noticed. The front line of the organization, where the actual actions of new directives have to take place, sees this as a constant change of direction. They start talking about the program of the month and, consequently, they do not change anything and the results of management efforts will be absent. If this goes on for some time, no sustainable results will be achieved.

In this situation, a long-term maintenance contract offers a possible solution. The contract has to be founded on the right principles and work processes, because, when these are not changed for a long period of time, your contractor can help eliminate the "lack-of-constancy-of-purpose phenomenon." With good leadership, the work processes and your results should continuously improve. It could be done without a contractor, but not in a system where each new facility manager or maintenance manager means a new program.

Finally, rather than outsourcing, consider "in-sourcing." Smaller organizations obviously have a greater difficulty allocating the needed to resources to building maintenance, finding and retaining skilled staff, etc. But, by combining several organizations' maintenance departments and implementing an adequate maintenance management program, maintenance can be improved and costs can be lowered. For example a small local school system may combine its maintenance operations with the local community college. In some communities, a larger university with a good maintenance program may be neighbors with one or more small colleges with less-effective maintenance programs.

COMPUTERIZED MAINTENANCE
MANAGEMENT SYSTEMS (CMMS)

A CMMS software package maintains a computer database of information about an organization's maintenance operations and, today, is the heart of an effective preventative maintenance program.

CMMS information is intended to help maintenance workers do

their jobs more effectively (for example, determining which storerooms contain the spare parts they need) and to help management make informed decisions (for example, calculating the cost of maintenance for each piece of equipment used by the organization, possibly leading to better allocation of resources). The information may also be useful when dealing with third parties; if, for example, an organization is involved in a liability case, the data in a CMMS database can serve as evidence that proper safety maintenance has been performed.

Different CMMS packages offer a wide range of capabilities and cover a correspondingly wide range of prices. A typical package deals with some or all of the following:

1. Work orders: Scheduling maintenance procedures, assigning personnel, reserving materials, recording costs, and tracking relevant information such as the cause of the problem (if any), downtime involved (if any), and recommendations for future action

2. Preventative maintenance (PM): Keeping track of PM components and procedures, including step-by-step instructions or check-lists, lists of materials required, and other pertinent details. Typically, the CMMS schedules PM jobs automatically based on required service and repair intervals.

3. Asset management: Recording data about equipment and property including specifications, warranty information, service contracts, spare parts, purchase date, expected lifetime, and anything else that might be of help to management or maintenance workers.

4. Inventory control: Management of spare parts, tools, and other materials including the reservation of materials for particular jobs, recording where materials are stored, determining when more materials should be purchased, tracking shipment receipts, and taking inventory.

CMMS packages can produce status reports and documents giving details or summaries of maintenance activities. The more sophisticated the package, the more analysis facilities are available.

Many CMMS packages can be either web-based, meaning they are hosted by the company selling the product on an outside server, or local, meaning that the user buys the software and hosts the product on his or her own server.

Following are the recommended minimum requirements for a CMMS for the management and maintenance of capital asset components only—they do not apply to unplanned maintenance efforts that are addressed through manual work order generation:

1. Database of preventative and predictive maintenance procedures for each facility component.
    a. Schedule of occurrence (daily, weekly, monthly, every X months, annually, every X years, etc.)
    b. Maintenance tasks, which include (1) significant inspection and diagnostic guides and required repair descriptions and/or (2) troubleshooting guidelines. These tasks shall be incorporated into a "master" library and then assigned to each element as required (see Appendices).

2. Work orders (WOs).
    a. WOs must be automatically generated on the basis of stipulated schedule of occurrence for maintenance tasks.
    b. Assign WOs to specific craft and/or staff member.
    c. WOs shall include required materials and tools list, safety requirements, and detailed maintenance tasks outlined in 1b above.
    d. WOs incorporate computer-based response by the assigned lead maintenance technician. Response shall include status or completeness of the WO, condition assessment of the component, and/or generation of required follow-up or follow-on work orders.

3. Maintenance support database is the data that are computer-resident in a form that will allow the maintenance staff to access the information as needed to perform a maintenance task and includes the following:
    a. System schematic drawings.
    b. Sequences of operation or control.
    c. Component data
        1) Component nameplate data, including manufacturer, model, serial number, date of installation.

2) Component drive motor data (if applicable) including rated horsepower, motor type, voltage/phase, amperage.
3) Capacity or other performance rating. (For example, fan capacity is rated as CFM at inches wg static pressure, pump capacity is rated as GPM at feet of water head, etc.)
4) Parts lists and any associated parts drawings (exploded views, etc.)
5) Manufacturer's detailed service and maintenance instructions.

4. The computerized preventative/programmed maintenance software should be (a) intuitive to use by a facilities or maintenance staff member and (b) reflect the best practices for facilities maintenance—*with less emphasis on "bean counting" or cost control of maintenance operations and more emphasis on performing the maintenance required to ensure building components achieve their design life.*

MAINTENANCE STAFF JOB DESCRIPTIONS

Facilities Maintenance Director

Responsible for directing and supervising the daily operations/ activities of the Facilities Maintenance Department, including the maintenance, repair, and upkeep of all facilities, and equipment, and developing, managing, and supervising budget, procurement, and contracting activities required to meet these responsibilities. Duties also include overall responsibility for the day-to-day operations of the maintenance department, establishing maintenance priorities, developing short- and long-term planning, managing the and operation of facilities service systems.

Minimum Education	Bachelor's Degree in Management, Engineering, Construction, Architecture, or a related field.
Minimum General Skill Set	1. Oversee the assignment of work orders and maintenance projects to crews. Provide directions as needed to crew members either directly or through the craft supervisors.
	2. Communicates with architects and engineers during the

design, modification, and inspection of existing and new facilities in regards to maintenance impact.

3. Plan, organize, coordinate, and supervise the maintenance, modification, and repair of buildings, facilities, and equipment.

4. Prepare and administer outside contracts for the required maintenance of facilities, including elevator, fire alarm, stand-by emergency power, HVAC controls, etc., as required.

5. Determination of maintenance priorities, including coordinating and scheduling projects, establishing work standards and operating procedures and preventative maintenance programs.

6. Prepares and approves purchase orders and accident reports.

7. Makes presentation of maintenance department's operations plan and budget to management.

8. Exercise full supervisory and decision making authority in the department, including the hiring, promoting, evaluating, and terminating personnel.

9 Initiate disciplinary action and other personnel-related activities as necessary.

Specific Skills Required	Have knowledge of: • personnel management and supervisory practices • occupational hazards and safety precautions of the building maintenance trade • building and safety codes • energy conservation methods and procedures • central plant and district heating/cooling systems • effective public relations methods and practices Have ability to: • participate in the establishment and implementation of work and safety standards

- maintain the department's records, prepare daily reports and control expenditures
- identify and resolve operational and personnel problems
- research and produce written documents, reports, and analysis with clearly organized thoughts and substantiated conclusions
- establish and maintain effective working relationships with other management, staff, contractors, customers and the general public

Minimum Experience	9+ years of progressively responsible managerial and administrative experience in a building and grounds maintenance, code enforcement, or facilities management operation, two years of which must have been in a supervisory capacity.

Craft Supervisor

Second or third tier management level in a maintenance department, responsible for directing and supervising the daily operations/activities of technicians with a specific craft area (general facility, HVAC, Plumbing, or Electrical). Plans, supervises, directs and reviews the work of skilled and semi-skilled and support staff performing journey-level installation, repair, and maintenance.

Minimum Education	2-year Associate Degree, completion of apprentice program, or military training in the a specific maintenance craft. 15 hours, minimum, each year of additional instruction in one or more craft areas.
Preferred Education	Bachelor's Degree in Management, Engineering, Construction, Architecture, or a related field.
Minimum General Skill Set	1. Plans, schedules, assigns, supervises, reviews and evaluates the work of skilled and semi-skilled staff performing maintenance and repair work. 2. Recommends selection of staff. 3. Trains staff in work procedures; administers discipline as required.

4. Assists in developing and implementing goals, objectives, procedures and work standards for specific craft areas and provides input into the maintenance budgets.

5. Troubleshoots problems and provides technical assistance to staff in solving maintenance and repair problems.

6. Requisitions materials, parts, tools and equipment needed for work assignments; orders and picks up materials and supplies from outside vendors as appropriate; prepares cost estimates for jobs.

7. Prepares, issues, prioritizes and/or monitors work orders.

8. Inspects repair and maintenance work in progress and upon completion to ensure that work complies with applicable codes and regulations.

9. Ensures that staff follows safety procedures and uses appropriate safety equipment.

Specific Skills Required	Have knowledge of: • craft area technical issues and maintenance procedures • personnel management and supervisory practices • occupational hazards and safety precautions of the building maintenance trade • building and safety codes
Minimum Experience	5+ years experience as Technician Level III, two years of which must have been in a supervisory capacity as foreman or lead technician.

General Facility Technician Level I

Entry or "apprentice" level maintenance worker who performs only under the direct and continuous supervision of a higher level technician.

Minimum Education	High school diploma, GED, or equivalent.
Desirable Education	2-year Associate Degree, completion of apprentice program, or military training in a specific maintenance craft area.

Minimum General Skill Set	Basic skill with hand tools and power tools normally utilized in a specific craft area.
	Knowledge of construction drafting and ability to read and understand blueprints.
	Working knowledge of workplace safety rules and procedures.
Craft Skill Set	Assist in servicing and replacing minor plumbing fixtures.
	Perform minor electrical work not requiring a licensed electrician such as replacing light bulbs and fuses.
	Assist in minor carpentry, masonry, roofing, and other general facility maintenance.
	Maintain machinery, equipment, and tools by cleaning, lubricating, greasing, oiling to ensure optimum working order.
Minimum Experience	1-2 years depending, on education, working in construction or building maintenance.

General Facility Technician Level II

Semi-skilled or "helper" level maintenance worker who performs under the general supervision of a higher level technician.

Minimum Education	2-year Associate Degree, completion of apprentice program, or military training in the a specific maintenance craft.
	15 hours, minimum, each year of additional instruction in one or more craft areas.
Minimum General Skill Set	All skills required of Level I technician.
	Knowledge and understand of building codes, facility standards, and applicable regulations for a specific craft.

	Capable of performance of basic preventative maintenance procedures within one or more craft areas with minimal supervision. Capable of training, supervising, and directing Level I technicians.
Craft Skill Set	Apply carpentry techniques and using carpentry tools to construct, alter, repair, and/or install walls, stairs, floors, ceiling, windows, doors, roofs, gutters, lock, etc. Perform structural repair, alterations, or improvement work such as preparing surfaces for plaster or cement work; studding, and plastering walls. Install glass in doors and windows.
Minimum Experience	3-5 years working in construction or building maintenance.

General Facility Technician Level III
Skilled or "journeyman" level maintenance worker.

Minimum Education	Same as for Level II technician
Minimum General Skill Set	All skills required of Level II technician. Capable of performance of basic preventative maintenance procedures within one or more craft areas with no supervision. Capable of performing inspection, analysis, and evaluation of facility components within one or more craft areas. Capable of directing and supervising Level I and II technicians.
Minimum Experience	6 or more years working in construction or building maintenance.

HVAC Technician Level I

Entry or "apprentice" level maintenance worker who performs only under the direct and continuous supervision of a higher level technician.

Minimum Education	High school diploma, GED, or equivalent.
Desirable Education	2-year Associate Degree, completion of apprentice program, or military training in a specific maintenance craft area.
Minimum General Skill Set	Basic skill with hand tools and power tools normally utilized in a specific craft area.
	Basic knowledge of shop math, mechanical drawing, applied physics and chemistry, electronics, blueprint reading, and computer applications.
	Some knowledge of plumbing and electrical work also is helpful.
	A basic understanding of electronics is required
	Working knowledge of workplace safety rules and procedures.
Craft Skill Set	Have a thorough grounding in the fundamentals of heating, ventilating, and air-conditioning system installation, operation, and troubleshooting for, at a minimum, residential and light commercial "comfort" applications.
Minimum Experience	1-2 years depending, on education, working in construction or building maintenance.

HVAC Technician Level II

Semi-skilled or "helper" level maintenance worker who performs under the general supervision of a higher level technician.

Minimum Education	2-year Associate Degree, completion of apprentice program, or military training in the a specific maintenance craft.

	15 hours, minimum, each year of additional instruction in one or more craft areas.
Minimum General Skill Set	All skills required of Level I technician. Have taken and passed technician certification test(s) at the "journeyman" level or higher. (The tests are offered through Refrigeration Service Engineers Society (RSES), HVAC Excellence, The Carbon Monoxide Safety Association (COSA), Air Conditioning and Refrigeration Safety Coalition, and North American Technician Excellence, Inc. (NATE), among others.) Be trained and hold an EPA-Approved Section 608 certification needed to service building air conditioning and refrigeration systems. Knowledge and understand of building codes, facility standards, and applicable regulations for a specific craft. Capable of performance of basic preventative maintenance procedures within one or more craft areas with minimal supervision. Capable of training, supervising, and directing Level I technicians.
Craft Skill Set	Have a basic grounding in the fundamentals of commercial, industrial, or institutional heating, ventilating, and air-conditioning system installation, operation, and troubleshooting. Capable of pipe and tube fitting (including measurement and layout), silver soldering, soft soldering, cutting/repairing steel pipe threads and assembling steel pipe and fittings, and welding (electric arc and MIG).
Minimum Experience	3-5 years working in construction or building maintenance.

HVAC Technician Level III
Skilled or "journeyman" level maintenance worker.

Minimum Education	Same as for Level II technician
Minimum General Skill Set	All skills required of Level II technician. Capable of performance of basic preventative maintenance procedures within one or more craft areas with no supervision. Capable of performing inspection, analysis, and evaluation of facility components within one or more craft areas. Capable of directing and supervising Level I and II technicians.
Craft Skill Set	Have a thorough grounding in the fundamentals of commercial, industrial, or institutional heating, ventilating, and air-conditioning system installation, operation, and troubleshooting. Capable of pipe and tube fitting (including measurement and layout), silver soldering, soft soldering, cutting/repairing steel pipe threads and assembling steel pipe and fittings, and welding (electric arc and MIG).
Minimum Experience	6 or more years working in construction or building maintenance.

Plumbing Technician Level I
Entry or "apprentice" level maintenance worker who performs only under the direct and continuous supervision of a higher level technician.

Minimum Education	High school diploma, GED, or equivalent.
Desirable Education	2-year Associate Degree, completion of apprentice program, or military training in a specific maintenance craft area.

Minimum General Skill Set	Basic skill with hand tools and power tools normally utilized in a specific craft area.
	Knowledge of construction drafting and ability to read and understand blueprints.
	Working knowledge of workplace safety rules and procedures.
Minimum Experience	1-2 years depending, on education, working in construction or building maintenance.

Plumbing Technician Level II

Semi-skilled or "helper" level maintenance worker who performs under the general supervision of a higher level technician.

Minimum Education	2-year Associate Degree, completion of apprentice program, or military training in the a specific maintenance craft.
	15 hours, minimum, each year of additional instruction in one or more craft areas.
Minimum General Skill Set	All skills required of Level I technician.
	Knowledge and understand of building codes, facility standards, and applicable regulations for a specific craft.
	Capable of performance of basic preventative maintenance procedures within one or more craft areas with minimal supervision.
	Capable of training, supervising, and directing Level I technicians.
Craft Skill Set	Review blueprints and building codes and specifications to determine work details and procedures.
	Study building plans and inspect structures to assess material and equipment needs, to establish the sequence of pipe installations, and to plan installation around obstructions such as electrical wiring.

	Assemble pipe sections, tubing and fittings, using couplings, clamps, screws, bolts, cement, plastic solvent, caulking, or soldering, brazing and welding equipment.
	Fill pipes or plumbing fixtures with water or air and observe pressure gauges to detect and locate leaks.
	Locate and mark the position of pipe installations, connections, passage holes, and fixtures in structures, using measuring instruments such as rulers and levels.
	Measure, cut, thread, and bend pipe to required angle, using hand and power tools or machines such as pipe cutters, pipe-threading machines, and pipe-bending machines.
	Test, repair, and replace plumbing fixtures, pumps, tanks, etc.
	Install pipe assemblies, fittings, valves, appliances such as dishwashers and water heaters, and fixtures such as sinks and toilets, using hand and power tools.
Minimum Experience	3-5 years working in construction or building maintenance.

Plumbing Technician Level III
Skilled or "journeyman" level maintenance worker.

Minimum Education	Same as for Level II technician
Minimum General	All skills required of Level II technician.
Skill Set	Capable of performance of basic preventative maintenance procedures within one or more craft areas with no supervision.
	Capable of performing inspection, analysis, and evaluation of facility components within one or more craft areas.
	Capable of directing and supervising Level I and II technicians.

Craft Skill Set	All skills required for Level II Technician
Minimum Experience	6 or more years working in construction or building maintenance.

Electrical Technician Level I

Entry or "apprentice" level maintenance worker who performs only under the direct and continuous supervision of a higher level technician.

Minimum Education	High school diploma, GED, or equivalent.
Desirable Education	2-year Associate Degree, completion of apprentice program, or military training in a specific maintenance craft area.
Minimum General Skill Set	Basic skill with hand tools and power tools normally utilized in a specific craft area. Knowledge of construction drafting and ability to read and understand blueprints. Working knowledge of craft and workplace safety rules and procedures, particularly NFPA *Standard 70E*.
Minimum Experience	1-2 years depending, on education, working in construction or building maintenance.

Electrical Technician Level II

Semi-skilled or "helper" level maintenance worker who performs under the general supervision of a higher level technician.

Minimum Education	2-year Associate Degree, completion of apprentice program, or military training in the a specific maintenance craft. 15 hours, minimum, each year of additional instruction in one or more craft areas.
Minimum General Skill Set	All skills required of Level I technician.

	Knowledge and understand of building codes, facility standards, and applicable regulations for a specific craft.
	Capable of performance of basic preventative maintenance procedures within one or more craft areas with minimal supervision.
	Capable of training, supervising, and directing Level I technicians.
Craft Skill Set	Knowledge of electrical power systems maintenance, repair and replacement, and installation of electrical power distribution raceway, wiring, devices, and equipment.
	Knowledge of the application and use of testing equipment for checking, testing, and analyzing electrical systems and equipment (including ammeter, oscilloscope, voltage tester, phase and rotation tester, and multimeter)
	Capable of sizing, routing, measurement, and installation of raceway systems, including EMT, REC, and FEC, along with junction boxes, device boxes, and panel/equipment connections.
	Capable of installing conductors in raceway systems... pulling wire and making connections/terminations.
	Must hold, a required, a state or local "journeyman electrician" or electrical contractor license.
Minimum Experience	3-5 years working in construction or building maintenance.

Electrical Technician Level III
Skilled or "journeyman" level maintenance worker.

Minimum Education	Same as for Level II technician
Minimum General Skill Set	All skills required of Level II technician.

	Capable of performance of basic preventative maintenance procedures within one or more craft areas with no supervision.
	Capable of performing inspection, analysis, and evaluation of facility components within one or more craft areas.
	Capable of directing and supervising Level I and II technicians.
Craft Skill Set	Knowledge of electrical power systems maintenance, repair and replacement, and installation of electrical power distribution raceway, wiring, devices, and equipment.
	Knowledge of the application and use of testing equipment for checking, testing, and analyzing electrical systems and equipment (including ammeter, oscilloscope, voltage tester, phase and rotation tester, and multimeter).
	Capable of sizing, routing, measurement, and installation of raceway systems, including EMT, REC, and FEC, along with junction boxes, device boxes, and panel/equipment connections.
	Capable of installing conductors in raceway systems... pulling wire and making connections/terminations.
	Maintain and repair low voltage systems including fire, alarm and access systems.
	Test electrical equipment for safety and efficiency.
Minimum Experience	6 or more years working in construction or building maintenance.

The following "specialty" maintenance staff require the highest level of skills and should, therefore, classified as Level III technicians.

Carpenter/Cabinetry Technician

Minimum Education	Same as for Level II technician
Minimum General Skill Set	All skills required of Level II technician.
	Capable of performance of basic preventative maintenance procedures within one or more craft areas with no supervision.
	Capable of performing inspection, analysis, and evaluation of facility components within one or more craft areas.
	Capable of directing and supervising Level I and II technicians.
Craft Skill Set	Capable of Maintaining buildings and related structures applying carpentry techniques and using carpentry tools to construct, alter, repair and/or install walls, stairs, floors, ceilings, windows, doors and door hardware, handicapped chair rails and ramps, shelving, etc.
	Inspect buildings needing carpentry repair of alterations and estimate job duration and the cost and/or quantities of labor and materials needed to complete the job.
	Interpret and/or draw up sketches, patterns, blueprints, instructions and/or layouts of work to be done.
	Perform structural repair, alteration or improvement work such as lowering ceilings and paneling rooms; preparing surfaces for plaster or cement work; studding, wiring, and plastering walls and cracks in floors or around windows; taping joints on GWB; and piece-matching and/or replacing countertops.
Minimum Experience	6 or more years working in construction or building maintenance.

Painter

Minimum Education	Same as for Level II technician
Minimum General Skill Set	All skills required of Level II technician.
	Capable of performance of basic preventative maintenance procedures within one or more craft areas with no supervision.
	Capable of performing inspection, analysis, and evaluation of facility components within one or more craft areas.
	Capable of directing and supervising Level I and II technicians.
Craft Skill Set	Apply paints, varnish, and/or stains to all types of surfaces (e.g. wood, metal, plaster, etc.) In order to protect and beautify such surfaces using appropriate utensils and equipment such as paint spraying machines.
	Prepare surfaces for painting using appropriate tools, equipment, and techniques.
	Select the types and colors of paints, varnish or stain and mixes, blends and prepares the same for proper color match.
	Prepare surfaces for plaster or cement work by taping joints on sheetrock, caulking and waterproofing walls and cracks in floors, etc. by using appropriate tools and equipment.
	Performs related duties such as removing and replacing glass; drilling holes in signs; printing letters and signs; and moving office furniture.
	Based on assignment, may operate road-line painting machines and sand blasting equipment.

Minimum Experience	6 or more years working in construction or building maintenance.
Locksmith	
Minimum Education	Same as for Level II technician
Minimum General Skill Set	All skills required of Level II technician. Capable of performance of basic preventative maintenance procedures within one or more craft areas with no supervision. Capable of performing inspection, analysis, and evaluation of facility components within one or more craft areas. Capable of directing and supervising Level I and II technicians.
Craft Skill Set	Install lock devices in doors, desks, office equipment and other units. Repair and/or overhaul locking devices such as mortise, rim, key-in-the knob locks, dead bolt, office equipment, padlocks and emergency exit locks by repairing for replacing worn tumblers, springs or other parts of locking devices; by making alterations to locking mechanisms and units; and/or by replacing escutcheons and face plates on doors to change the size of existing holes. Install locks containing dual locking capabilities by reworking locking devices designed to work in conjunction with other components and equipment to form a complete security of surveillance systems. Changes lock combination by inserting new or repaired tumblers into locks in order.
Minimum Experience	4+ years of full-time experience in locksmithing.

Conveying Systems Technician

Minimum Education Minimum General Skill Set	Same as for Level II technician All skills required of Level II technician. Capable of performance of basic preventative maintenance procedures within one or more craft areas with no supervision. Capable of performing inspection, analysis, and evaluation of facility components within one or more craft areas. Capable of directing and supervising Level I and II technicians.
Craft Skill Set	Knowledge of the current editions of the following American Society of Mechanical Engineers (ASME) elevator codes: ASME A17.1, *Safety Code for Elevators and Escalators* ASME A17.2, *Inspectors' Manual for Elevators and Escalators* ASME A17.3, *Safety Code for Existing Elevators and Escalators* ASME A17.4, *Guide for Emergency Evacuation of Passengers from Elevators* ASME A17.1 *Handbook* and ASME A17.1 Interpretations ASME QEI-1 Certification as a Qualified Elevator Inspector, as defined by ASME QEI-1. Inspect and test elevators and related machinery to ensure that they meet legal operating requirements.

	Coordinate and schedule inspections with other agencies and contractors.
Minimum Experience	6 or more years working in construction or building maintenance.

Masonry/Tile Technician

Minimum Education	Same as for Level II technician
Minimum General Skill Set	All skills required of Level II technician. Capable of performance of basic preventative maintenance procedures within one or more craft areas with no supervision. Capable of performing inspection, analysis, and evaluation of facility components within one or more craft areas. Capable of directing and supervising Level I and II technicians.
Craft Skill Set	Lay brick, stone, and CMU and other masonry units using a variety of hand and power tools such as brick or napping hammer, brick or shill saw, pick, chisel and hammer to build, alter, repair and/or maintain brick and/or concrete surfaces and structures such as sidewalks, partitions, floors, chimneys, etc. Repair or install ceramic time, stone, or brick flooring. Perform related duties such as caulking and waterproofing cracks in masonry walls and floors and coring holes through concrete or masonry walls or floors.
Minimum Experience	6 or more years working in construction or building maintenance.

Roofing Technician

Minimum Education Minimum General Skill Set	Same as for Level II technician All skills required of Level II technician.
	Capable of performance of basic preventative maintenance procedures within one or more craft areas with no supervision.
	Capable of performing inspection, analysis, and evaluation of facility components within one or more craft areas.
	Capable of directing and supervising Level I and II technicians.
Craft Skill Set	Maintain, repair, and replace all types of roofing materials using tools and equipment of the trade.
	Repairs leaks in metal roof, walls, and ceilings by soldering, crimping, riveting, and welding seams in sheet metal and installing flashing and waterproof felt using appropriate hand and power tools to reduce water damage and restore water-tight integrity.
	Fabricate miscellaneous metal parts such as gavel guards, gutters, spouts, pipes, by cutting, drilling, tapping and soldering metal to meet roofing requirements.
Minimum Experience	6 or more years working in construction or building maintenance.

HVAC Controls Technician

Minimum Education	Same as for Level II technician
Minimum General Skill Set	All skills required of Level II technician.

	Capable of performance of basic preventative maintenance procedures within one or more craft areas with no supervision. Capable of performing inspection, analysis, and evaluation of facility components within one or more craft areas. Capable of directing and supervising Level I and II technicians.
Minimum Experience	6 or more years working in construction or building maintenance.

Fire Sprinkler Technician

Minimum Education	Same as for Level II technician
Minimum General Skill Set	All skills required of Level II technician. Capable of performance of basic preventative maintenance procedures within one or more craft areas with no supervision. Capable of performing inspection, analysis, and evaluation of facility components within one or more craft areas. Capable of directing and supervising Level I and II technicians.
Craft Skill Set	Knowledge of NFPA *Standard 25*, "Standard for the Inspection, Testing, and Maintenance of Water-Based Fire Protection Systems." In many locales, the personnel performing these functions must be licensed by the state or local government, as follows: Personnel performing <u>maintenance</u> on fire sprinkler systems must hold a valid Limited *Fire Sprinkler System Maintenance Technician* license.

	Personnel performing *testing and inspection* of fire sprinkler systems must hold a valid *Limited Fire Sprinkler System Inspection Technician* license.
Minimum Experience	6 or more years working in construction or building maintenance.

Special Electrical Systems Technician

Dedicated to fire alarm, security, communications, information technology, etc.

Minimum Education	Same as for Level II technician
Minimum General Skill Set	All skills required of Level II technician. Capable of performance of basic preventative maintenance procedures within one or more craft areas with no supervision. Capable of performing inspection, analysis, and evaluation of facility components within one or more craft areas. Capable of directing and supervising Level I and II technicians.
Craft Skill Set	Working knowledge and significant experience with one or more special electrical systems, including fire alarm, security, communications, information technology, etc.
Minimum Experience	6 or more years working in construction or building maintenance.

Chapter 2

Planned Replacement or Major Renovation

The concept of entropy is that nature tends from order to disorder in isolated systems—therefore, component failure is inevitable.

BUILDING COMPONENT LIFE

Every building is assembled from a great many individual components—materials, systems, sub-systems, devices, equipment, etc. But, *every* one of these building components will ultimately fail and must be replaced or renovated when it does. Historically, the life-cycle of building components has been represented by the "bathtub curve" of Figure 2-1.

However, the bathtub curve does not define the point at which "normal aging" failures finally require that the component be replaced or renovated.

Maximum "performance life" of a building component is the time over which the component serves its anticipated function over the range from 100% (when installed or initially placed in service) to 0% (when it fails and the only option is to replace it). However, once the level of performance falls below some minimum (just how many roof leaks are acceptable?) *and* the cost of continuing to maintain a failing component threatens to exceed the cost of replacement, then the component has reached the end of its "design service life" (often referred to as its "economic life").

Figure 2-2, an enhancement of the "bathtub curve," ignores "infant mortality" failures associated with new construction and shows a more definitive relationship between component performance and time. Over any component's performance life, function degrades slowly, even with adequate maintenance, until a "tipping point" is reached where function begins to degrade at an accelerating rate until failure occurs.

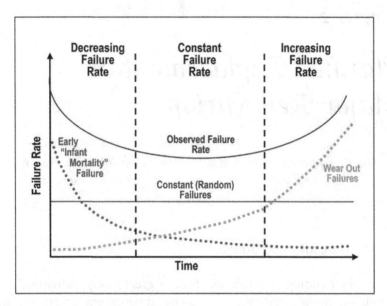

Figure 2-1. Component Failure as a Function of Time (Source: NASA)

The cost of maintenance follows an inverse relationship, as shown in Figure 2-3. While routine and preventative maintenance is a relatively small (but constant) requirement from day one, as the tipping point is approached, maintenance problems and costs begin to accelerate dramatically until final failure occurs and continued use and maintenance of the component is no longer an option.

However, if replacement is delayed until final failure occurs, the total cost of maintenance will inevitably exceed the cost of replacement that would have been incurred as the tipping point was reached at the end of the design service life. Design service life does not imply that there is no performance life left in the component, but it does imply that the cost of obtaining that remaining performance will exceed the cost of replacement at that point.

The best current estimates for the "reference" or median service life of building components are presented in *Component Life Manual*, Housing Association Property Mutual Ltd. (HAPM), London, UK (Published by E. & F.N. Spon, Andover, Hants, UK.), 1999. However, the data contained in this publication are generally limited to materials commonly applied to housing and light commercial construction in the United Kingdom and are somewhat incomplete when compared to con-

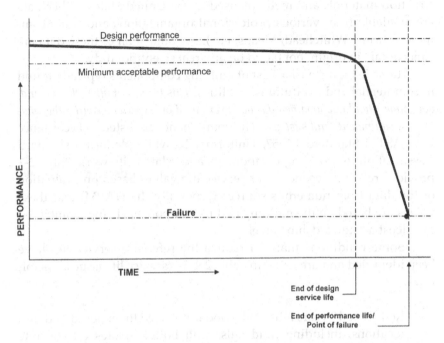

Figure 2-2. Component Performance as a Function of Time

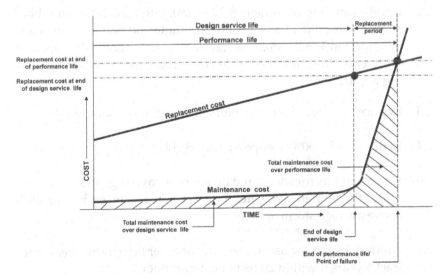

Figure 2-3. Maintenance/Replacement Cost as a Function of Time

struction materials and methods used in the United States. Other data are available from various professional organizations and some commercial firms (Whitehead, Means, etc.), but almost all of these data are "anecdotal" in origin, not based on any real scientific study.

Table 2-1 is a "master" list of building components typically found in commercial and institutional facilities. *This list is not all-inclusive and, over time, should be modified by each user based on experience with other materials, equipment, and systems.* The components are listed in accordance with ASTM Standard E1557, Uniformat II, which defines a standard classification for building components and related site work. Table 2-1 provides reference economic or service life values based on evaluation of the data from numerous sources (especially the HAPM) and these values are based on the building and its components being maintained to at least a "stewardship" level.

Some conditions that can reduce the reference service life have been identified and are listed in Table 2-1 as "service life issues," as follows:

2.1   Reduce to 10 years if treated wood is not used for exposed or damp locations, including mud sills, wall bottom plates on concrete slabs, the first row of exterior sheathing, etc.

2.2   Acrylic caulking is average (8-12 years). Butyl and bitumen rubber caulks have a life of 5 years. Polysulphide and polyurethane have an expected life of 15-20 years. Backer rods should be replaced with recaulking. Subtract 5 years for second and subsequent caulking applications.

2.3   Reduce to 5 years for dark colors facing south or southwest.

2.4   Reduce life by 50% if slope exceeds 3/12.

2.5   Storms and hurricanes can damage roof coverings enough to significantly shorten their life. Subtract 2 years for each year with above average storm activity.

3.1   Subtract 3 years for use in corridors or other high traffic areas. Subtract 5 years if within 20 feet of exterior door.

3.2 High humidity levels, roof leaks, or other moisture events will require ceiling replacement.

3.3 Vinyl wall covering installed on an exterior wall that has no outboard vapor retarder can create a mold infestation condition, requiring immediate removal of the wall covering.

3.4 Excess building movement due to thermal expansion/contraction, wind, or differential settlement can cause sealants to separate, requiring immediate replacement.

3.5 Physical damage by carts, cleaning operations, etc. can reduce life by as much as 50%.

4.1 Reduce life by 40% for outdoor application. Outdoor installation requires a second, weatherproof covering to protect the insulation.

4.2 In coastal areas, condenser fins and unit casing must have epoxy or phenolic coating to protect aluminum from rapid oxidation. Without this coating, life is reduced by 70%.

4.3 Plumbing fixtures in middle schools suffer greater vandalism. Reduce life by 20% in middle schools.

6.1 Reduce life by 3 years each time the classroom is relocated.

Table 2-1. Typical Component Service Life

Level 1 Major Group Elements	Level 2 Group Elements	Level 3 Individual Elements	Level 4 Sub-Elements	Service Life (yrs)	Service Life Issues
1. SUB-STRUCTURE	1. Foundations	1. Standard Foundations	Wall foundations	60+	None
			Column foundations	60+	None
			Foundation drainage	60+	None
		2. Special Foundations	Pilings/pile caps	60+	None
			Grade beams	60+	None
			Caissons	60+	None
			Overburden	60+	None
	2. Slab on Grade		Standard slab on grade	60+	None
			Structural slab on grade	60+	None
			French drains	35	None
			Perimeter insulation	35	None

Table 2-1. Typical Component Service Life (*Continued*)

Level 1 Major Group Elements	Level 2 Group Elements	Level 3 Individual Elements	Level 4 Sub-Elements	Service Life (yrs)	Service Life Issues
		3. Basement Walls	Basement wall construction	60+	None
			Wall moisture protection	35	None
2. SHELL	1. Super-structure	1. Floor Construction	Structure: Wood (Pine, fir, etc.)	35+	2.1
			Laminated wood (LVL)	25-30	None
			Masonry	60+	None
			Structural steel frame	60+	None
			Concrete frame	60+	None
			Floor deck: Wood	35+	2.1
			Concrete	60+	None
			Floor insulation	35+	None
			Balcony floors: Cast-in-place concrete	25	None
			Precast concrete	10-15	None
			Exterior stairs/fire escapes:		
			Steel, painted or galvanized	15	None
			Steel, hot-dipped galvanized	25	None
			Stainless steel	35+	None
		2. Roof Construction	Flat deck: Plywood, OSB	25	None
			Metal	35+	None
			Metal w/concrete	35+	None
			Roof insulation	Roof life	None
			Pitched deck: Plywood, OSB	30	None
			Metal	35+	None
			Metal w/concrete	35+	None
			Roof insulation	35+	None
			Facia/Soffits/Coping:		None
			Wood	15	None
			Aluminum (coated)	20	None
			Steel, painted, galvanized	15	None
			Steel, stainless	25	None
			Vinyl	15	None
			Fiberglass panels	25	None
			Natural stone	30	None
			Cast stone	35	None
			Precast concrete	35	None
			Cement asbestos	25	None
			Flashing: Galv. metal	20	None
			Stainless steel	30	None
			Copper	30	None
			Flat roof edging: Alum. parapet cap	20	None
			Alum. gravel stop	20	None
			Expansion joint: Prefabricated	20	None
			Roof joint w/alum.	20	None
			Ridge vent: Alum	23	None
			Molded polyethylene	20	None
			Canopies: Aluminum	20	None
			Canvas awnings	7	None
			Flat roof drains	35+	None
			Flat roof scuppers	20	None
			Gutters/downspouts: Aluminum	20	None
			Galv. steel	20	None
			Stainless steel	30	None
			Copper	30	None
			PVC	5-7	None

Table 2-1. Typical Component Service Life (*Continued*)

Level 1 Major Group Elements	Level 2 Group Elements	Level 3 Individual Elements	Level 4 Sub-Elements	Service Life (yrs)	Service Life Issues
	2. Exterior Closure	1. Exterior Walls	Exterior walls: Precast concrete	42	None
			Brick veneer	60+	None
			CMU	35+	None
			EFIS (over masonry)	21	None
			Alum/vinyl siding	25	None
			Wood siding	20	None
			Cement board siding	35	None
			Cement stucco	35+	None
			Metal building panels	25	None
			Caulking (all)	10.5	2.2
			Flashing: Sheet metal	20	None
			Lintels: Painted steel	20	None
			Galvanized steel	25	None
			Stainless steel	60+	None
			Concrete	60+	None
			Exterior paint on wood or metals	7-10	2.3
		2. Exterior Windows	Windows: Steel windows	15	None
			Vinyl casement windows	19	None
			Vinyl sash windows	17	None
			Aluminum windows	23	None
			Wood windows	20	2.3
			Curtainwalls and Storefronts:		
			Steel	15	None
			Aluminum (Anodized)	35	None
			Window blinds	7	None
		3. Exterior Doors	Exterior doors: Hollow metal	15	None
			Solid core, wood	20	None
			Aluminum	20	None
			Fiberglass	25	None
			Door hardware: Automatic openers	7	None
			Closer	12	None
			Lockset	15	None
			Panic device	7	None
			Door stop	25	None
			Overhead doors: Wood	25	None
			Steel	15	None
			Aluminum	25	None
			Fiberglass	35+	None
			Electric operator	20	None
			Coil doors: Steel	15	None
			Electric operator	20	None
			Air curtains (Cafeteria/kitchen)	5-7	None
	3. Roofing	1. Roof Coverings	Pitched roofing: Asphalt shingles	15	2.5
			Asphalt rolled	5-10	2.5
			Wood shingles	15-25	2.5
			Slate tiles	60+	2.5
			Metal	25-35	2.5
			Built-up	15	2.4, 2.5
			Flat roofing: EPDM	10-12	2.5
			Other single membranes	20	2.5
			BUR, 3-ply	17	2.5
			BUR, 5-ply	25-30	2.5
		2. Roof Openings	Skylights	30	2.5
			Roof hatches	35+	2.5
			Roof ventilators	25	2.5
			Equipment curbs/penetrations	30	2.5

Table 2-1. Typical Component Service Life (*Continued*)

Level 1 Major Group Elements	Level 2 Group Elements	Level 3 Individual Elements	Level 4 Sub-Elements	Service Life (yrs)	Service Life Issues
3. INTERIORS	1. Interior Construction	1. Partitions	Fixed partitions: Metal studs/GWB	30	None
			Wood studs/GWB	30	None
			Plaster on lathe	35+	None
			CMU	60+	None
			Moveable partitions	10-15	None
			Toilet partitions: Metal	15	None
			Fiberglass	15	None
			Stone	30+	None
			Interior glass and storefronts	35+	None
		2. Ceilings	Acoustical ceiling: Lay-in	35	3.1,3.2
			Concealed spline	20	3.1,3.2
			Hard ceiling: GWB on steel runners	40+	3.2
			Plaster on lathe	60+	3.2
		3. Interior Doors	Doors: Wood, hollow	5	None
			Wood, solid	15	None
			Hollow metal	20	None
			Fiberglass	25	None
			Door hardware: Automatic openers	7	None
			Closer	12	None
			Lockset	15	None
			Panic device	7	None
			Door stop	25	None
		4. Specialties	Storage shelving/cabinets/lockers	10-15	None
			Ornamental metals and handrails	20	None
			Signage	10	None
			Closet specialties	15	None
			White boards	15	None
			AV equipment brackets/stands	10	None
			Lavatory counters	10	None
	2. Staircases	1. Stair Construction	Stair construction: Steel/concrete	35+	None
			Stair handrails and balustrades: Steel	35+	None
		2. Stair Finishes	Tread and landing finishes	10	None
			Handrail and balustrade finishes	7	None
	3. Interior Finishes	1. Wall Finishes	Painting: Corridors/high traffic areas	3	3.1
			Classrooms	5	None
			Offices, conference	7	None
			Storage rooms, MER, etc.	10	None
			Vinyl wall covering	10	3.3
			Ceramic tile: Mortar laid	35+	None
			Adhesive backing	17	None
			Caulking	3-10	3.4
		2. Floor Finishes	Floor coverings: Carpeting, carpet tile	11	3.1
			Vinyl tile	10	3.1
			Vinyl sheet	15	3.1
			VCT	22	3.1
			Terrazzo	35+	None
			Wood, solid, T&G	20	3.1
			Wood, laminated	12-15	3.1
			Epoxy	17	3.1
			Painting, Conc/wood	5-7	3.1
			Brick/terra cotta tile	60+	None
			Ceramic tile	20	None
			Bases, curbs, trim	15-20	3.5

Table 2-1. Typical Component Service Life (*Continued*)

Level 1 Major Group Elements	Level 2 Group Elements	Level 3 Individual Elements	Level 4 Sub-Elements		Service Life (yrs)	Service Life Issues
4. SERVICE SYSTEMS	1. Conveying Systems	1. Elevators, escalators, dumbwaiters, etc.			35+	None
		2. Wheelchair lifts			15	None
	2. Plumbing	1. Plumbing Fixtures and Equipment	Water closets		35	4.3
			Urinals		35	4.3
			Lavatories		25	4.3
			Sinks		25	4.3
			Showers		25	4.3
			Faucets/brass: Commercial		15	4.3
			Drinking fountains and coolers		15	4.3
			Kitchen fixtures		25	None
			Dishwashers		15	None
			Grease/art traps		7-10	None
			Water heaters: Gas/oil/elec ≤120 gal		15	None
				Gas/oil/elec >120 gal	20	None
				Steam/storage	20	None
				Steam/instantaneous	25	None
		2. Domestic Water Distribution	Copper water piping/valves		35+	None
			Plastic water piping/valves		30	None
			Pipe insulation		30	None
			Pumps, line-mounted		10	None
			Backflow preventer (RPP/RPZ type)		15	None
		3. Sanitary Waste	Waste piping		35+	None
			Vent piping		25	None
			Floor drains		25	None
			Pipe insulation		30	None
			Pumps/lift station		10	None
		4. Rain Water Drainage	Piping (Interior)		35+	None
			Piping (Exterior)		15	None
			Pipe insulation		30	None
		5. Special Plumbing Systems	Gas piping		35+	None
			Fuel oil piping		20	None
			Acid waste systems		25	None
			Pool piping and equipment		15	None
	3. HVAC	1. Air-conditioning unit, packaged or split system	Window unit		7	4.2
			Packaged or split system, <10 tons		15	4.2
				≥10 tons	17	4.2
			PTAC or PTHP		10	4.2
			Ductless split system		15	4.2
			Computer room unit		15	4.2
		2. Heat pumps	Packaged or split system, < 7.5 tons		15	4.2
				≥10 tons	17	4.2
			Water source units		19	None
		3. Air-handling units	Indoor		25	None
			Outdoor, rooftop		15	None
			Fan coil units		20	None
		4. Boilers	Steel firetube, steam		25	None
			Cast iron, steam or hot water		25	None
			Electric, hot water		15	None
			Pressure burner, gas, oil, or dual fuel		21	None
			Atmospheric burner, gas		12	None

Table 2-1. Typical Component Service Life (*Continued*)

Level 1 Major Group Elements	Level 2 Group Elements	Level 3 Individual Elements	Level 4 Sub-Elements	Service Life (yrs)	Service Life Issues
		5. Furnaces	Gas- or oil-fired	18	None
		6. Unit heaters	Gas- or oil-fired	13	None
			Hot water or steam	20	None
			Electric	10	None
		7. Radiant heaters	Gas-fired infrared heaters	15	None
		8. Air Terminals	CV, VAV boxes with or without reheat	20	None
			Fan-powered boxes	17	None
		9. Ductwork	Duct work	30	None
			Diffusers, grilles, and registers	25	None
			Humidifiers	15	None
		10. Fans (other than part of air-handling units)	Centrifugal	25	None
			Axial	20	None
			Propeller	15	None
			Exhaust, roof- or sidewall-mounted	20	None
		11. Coils (other than part of air-handing units)	DX, water, or steam	20	None
			Electric	15	None
		12. Heat Exchangers	Shell-and-tube, steam or HW	24	None
			Plate and frame	20	None
		13. Chillers	Air-cooled, electric-drive	15-17	4.2
			Water-cooled, electric drive	23	None
			Absorption, 2-stage direct-fired	15	None
		14. Cooling towers	Galvanized	17-20	None
			Galvanized w/stainless basin/deck	25	None
			Stainless steel	30+	None
			Ceramic	34	None
		15. Condensers	Air-cooled	20	4.2
			Evaporative	20	4.2
		16. Pumps	Base-mounted	20	None
			Line-mounted	10-15	None
			Sump pump	10	None
			Condensate pump, indoor, above grade	15	None
			Condensate pump, outdoor, below grade	10-12	None
		17. Controls	Pneumatic	20	None
			Electric	16	None
			DDC	8-12	None
		18. Piping, valves, and specialties	Water piping and valves	35+	None
			Steam piping and valves	35+	None
			Condensate piping and valves	15	None

Table 2-1. Typical Component Service Life (*Continued*)

Level 1 Major Group Elements	Level 2 Group Elements	Level 3 Individual Elements	Level 4 Sub-Elements	Service Life (yrs)	Service Life Issues
			Water specialties: Expansion tanks	20	None
			Flex connectors	15	None
			Strainers	25	None
			Air separators	25	None
			Pres. relief valves	10	None
			BFP (RPP/RPZ)	15	None
			BFP (Double chck)	15	None
			Steam specialties: PRV's	15	None
			Pres. relief valves	10	None
		19. Insulation, pipe/duct	Fiberglass	15	4.1
			Calcium silicate	15	4.1
			Celluar glass	25	4.1
		20. HVAC electric motors		15-30	None
	4. Fire Protection	1. Sprinkler Systems	Sprinkler water supply	35+	None
			Piping	35	None
			Air source for dry pipe systems	25	None
			Fire pump	35	None
		2. Standpipe & Hose Systems	Standpipe water supply	35+	None
			Piping	35	None
			Fire hose/cabinets	15	None
		3. Fire Protection Specialties	Fire extinguishers/cabinets	15	None
			Kitchen hood fire suppression system	15	None
			Special fire protection systems	15	None
	5.Electrical	1. Electrical Service & Distribution	High voltage service and distribution	25	None
			Low voltage service and distribution	25	None
		2. Lighting & Branch Wiring	Branch circuit wiring	35+	None
			Panels/breakers	25	None
			Wiring devices	15	None
			Light fixtures	15	None
		3. Communication, Fire Alarm, and Security Systems	Public address system	10-15	None
			Intercommunication/paging system	10-15	None
			Call systems	10-15	None
			Clock systems	10-15	None
			Fire alarm system	10-15	None
			Security monitoring and alarm systems	10	None
			CCTV systems	8	None
		4. Special Electrical Systems	Grounding	35+	None
			Emergency lighting	12	None
			Emergency power systems	12	None
			Static uninterruptible power systems	12	None
			Rotary uninterruptible power systems	12	None
5. EQUIPMENT & FURNISHINGS	1. Equipment	1. Commercial Equipment	Office equipment/computers	5	None
		2. Institutional Equipment	Library equipment	10	None
			Theater and stage equipment	15	None
			AV equipment	7	None
			Laboratory equipment	10	None
			Art equipment	10	None
		3. Vehicular Equipment	Parking control/gates	10	None
			Loading dock levelers, etc.	15	None

Table 2-1. Typical Component Service Life (*Continued*)

Level 1 Major Group Elements	Level 2 Group Elements	Level 3 Individual Elements	Level 4 Sub-Elements	Service Life (yrs)	Service Life Issues
		4. Other Equipment	Food service coolers/freezers	17	None
			Kitchen hood	20	None
	2.Furnishings	1. Fixed Furnishings	Fixed casework	20	None
			Blinds and other window treatment	7	None
			Seating	10	None
			Gymnasium flooring/removable wood	10	None
6 SPECIAL CONSTRUCTION	1. Special Structures		Pre-engineered metal buildings	25+	None
			Portable classrooms	12	None
	2. Special Facilities		Aquatic facilities	15	None
			Fuel oil storage tanks	25+	6.1
7. BUILDING SITEWORK	1. Site Improvements	1. Roadways	Bases/sub-bases	35+	None
			Paving and surfacing:	5-7	None
			Gravel/shellrock	30	None
			Concrete Asphalt	16	None
			Curbs, gutters, and drains	35+	None
			Pavement lines and markings	5-10	None
			Signage	12	None
			Bridges	12 35+	None
		2. Parking Lots	Bases/sub-bases	35+	None
			Paving and surfacing:	5-7	None
			Gravel/shellrock	30	None
			Concrete	16	None
			Asphalt	35+	None
			Curbs, gutters, and drains	5-7	None
			Pavement lines and markings	12	None
			Signage	12	None
		3. Pedestrian Paving	Concrete paving	25	None
			Asphalt paving	15	None
			Steps (concrete)	15	None
			Pavement lines and markings	5-7	None
			Signage	12	None
			Bridges	10-35+	None
		4. Site Development	Fencing/gates	10-15	None
			Retaining walls	15	None
			Signage	10	None
			Site furnishings	10	None
			Fountains, pools, water features	10	None
			Flagpoles	20+	None
			Playgrounds and play equipment	15	None
		5. Landscaping	Erosion/runoff control features	15	None
			Irrigation systems	10	None
	2. Site Civil / Mechanical Utilities	1. Water Supply & Distribution Systems	Well water	20-30	None
			Well pump: Submersible	15-22	None
			Above ground	13-20	None
			Piping and valves	20+	None
			Pressure tank: Steel	13-25	None
			Fiberglass	25-50	None
			Backflow preventer (RPP/RPZ)	15	None
			Water softener	20-30	None

Table 2-1. Typical Component Service Life (*Continued*)

Level 1 Major Group Elements	Level 2 Group Elements	Level 3 Individual Elements	Level 4 Sub-Elements	Service Life (yrs)	Service Life Issues
		2. Sanitary Sewer Systems	Piping	35+	None
			Manholes and cleanouts	35+	None
			Septic tank: Concrete	25-30	None
			Septic field	15	None
			Lift station	10	None
			Package waste treatment system	15	None
		3. Storm Sewer Systems	Piping	35+	None
			Manholes and cleanouts	35+	None
			Headwalls/catch basins	35+	None
			Ditches and culverts	20	None
			Retention ponds	20	None
			Lift station	10	None
		4. Heating Distribution	Piping and valves: Hot water	35	None
			Steam	20	None
			Condensate	12	None
			Manholes	35+	None
			Exposed insulation w/weather jacket	10	None
		5. Cooling Distribution	Piping and valves: CDW, CHW	35+	None
			Exposed insulation w/weather jacket	10	None
		6. Fuel Distribution	Gas piping: Above grade, painted	25	None
			Below grade, protected	30	None
			Below grade, plastic	40	None
			Oil piping: Above grade, painted	25	None
			Below grade, protected	30	None
			Below grade, plastic	40	None
	3. Site Electrical Utilities	1. Exterior Lighting	Fixtures, poles, and stanchions	15	None
			Raceway and wiring	25	None
			Lighting controls	10	None
		2. Exterior Communications & Security	Public address system	10	None
			Intercommunication/paging system	10	None
			Security monitoring and alarm systems	10	None
			CCTV systems	5-7	None

FUNDED DEPRECIATION AND BUDGETING
FOR REPLACEMENT/RENOVATION

Planned replacement or renovation of a building component when it fails is a major part of any effective programmed maintenance effort.

For many building owners, the approach to funding needed replacement or renovation is to budget funds as "capital investments" to address the need, but only after the need arises. The impact of these sudden, significant capital needs on annual budgets is typically so great, that only the most well-heeled owner can make funds available when needed. Most owners attempt to defer the cost as long as possible resulting in deferred maintenance and, ultimately, even greater cost (see Life Cycle Cost Analysis at Replacement/Renovation Time, below).

Sophisticated building owners incorporate a funding plan as part of planned replacement or renovation of building components. This plan, in accounting language, is referred to as funded depreciation, but a more accurate term is reinvestment or recapitalization. Under this scheme, funds are set aside each year, usually in safe, interest-bearing accounts, so that when a component reaches the end of its service life, the capital required for its replacement or renovation is already allocated and available.

The first step in this process is to establish each component's service life.

For initial planning for replacement/renovation of components in existing facilities, the referenced service life from Table 2-1 can be used as the starting point to determine the anticipated "remaining design service life" for each component. Longer service life can generally be obtained from better materials, designers having exercised better quality control, better construction quality, and more comprehensive maintenance. Likewise, poorer efforts in these areas, along with more severe environmental conditions, heavy "wear-and-tear" usage, and/or poor maintenance will certainly shorten service life. Owners must adjust the service life used for analysis for these various factors.

Two technical standards exist that define methodologies for evaluating the service life of buildings or building components:

- ISO Standard 15686-1, *Building and Construction Assets Service Life Planning: General Principles*

- Canadian Standards Association (CSA) S478-95 (R2007), *Guideline on Durability in Buildings*

Both standards are based on the following approach for defining factors that influence remaining design service life of building components:

$$RDSL = RSL \times A \times B \times C \times D \times E \times F$$

Where
RDSL = estimated (or anticipated) "remaining design service life"
RSL = reference service life provided in Table 2-1
A = quality of materials factor (poor, fair, good, better, best)

B = design quality control factor (poor, fair, good, better, best)

C = construction quality control factor (poor, fair, good, better, best)

D = environmental conditions factor (non-typical environmental conditions such as a harsh coastal climate or one with long cold winters, solar radiation, hot-humid climates, urban environments, pollutants, etc.)

E = in-use conditions factor (high level of wear-and-tear, such as K12 schools, building public areas, and parking areas or abuse, such as correctional facilities)

F = maintenance level factor

Each of factors A through F must be evaluated and an initial value between 0.8 and 1.2 assigned to each.

As part of the preventative maintenance program detailed in Chapter 3, each building component must be inspected and evaluated on a periodic, routine basis. Based on these evaluations, the RDSL for each component must then be adjusted, at least annually. (Note that the RDSL can be a negative number if needed replacement or major renovation has been postponed or "deferred.")

This process, then, is a systematic approach to answering the "repair or replace" question for aging building components. During this process, the answers to the following questions are being determined:

1. What is the age of the component relative to its design service life? If the age is well within the design service life, then the tendency would be to consider repair. If the RDSL is low, then the tendency would be to replace.

2. What is the current condition indicated by condition appraisals made during preventative maintenance procedures? Look at condition trends...condition degradation and increased repairs may indicate that replacement time has come.

3. Are spare parts still available? In areas of rapid technological change, such as building DDC systems, fire alarm systems, etc., repair is not feasible if spare parts are not available. This may dictate replacement even when the condition of the component is still good.

4.   Would replacement reduce maintenance costs, reduce energy costs, and/or reduce water costs to an extent that an acceptable payback period can be achieved? If so, the cost of replacement may be justified.

The best approach to budgeting for replacement/renovation maintenance is to apply the concept of *funded depreciation* to each building component. This concept is often called "re-capitalization" or "reinvestment" (though it's known historically as a "sinking fund") and it requires building owners to set aside funds each year, usually placed in an interest-bearing investment, that will be available in the future to pay to the cost of major replacement or renovation of a building component when that component reaches the end of its RDSL. This approach, adopted by many industrial firms and even some state governments (such as Minnesota), ultimately eliminates the need for borrowed funds (loans or bonds) and corresponding debt service obligations. And, as shown by the following discussion, *this concept eliminates deferred maintenance and lowers the costs of replacement/renovation maintenance.*

Using this concept, the current cost of replacement or renovation of each component, y, is determined, along with an estimate of its RDSL. Dividing the replacement/renovation cost by the number of years of RDSL yields the budget amount that must be allocated each year between now and the time of the replacement/renovation.

Future replacement or renovations costs will be higher than current costs due to inflation and, thus, a *cost escalation factor (CEF)* must be incorporated into the calculation. Inflation has a significant affect on future costs and under estimating the impact of inflation is the "kiss of death" for adequate funded depreciation.

Construction cost inflation is almost always greater than the general inflation rate, as illustrated by the data in Table 2-2 comparing general inflation and construction cost inflation rate in the United States over each decade since 1970.

Thus, it is good planning to assume that construction cost inflation will be 30-50% higher than the general inflation rate, a condition that has existed for the last (almost) 40 years. Otherwise, the funds available at the end of the component's service life to pay for replacement/renovation could be woefully inadequate.

For school boards and county commissioners, the "sticker shock" associated with needed replacements or renovations is a key cause of deferred maintenance—a roof replacement that would cost $1,000,000 today

Table 2-2. Inflation Rate Comparison

Decade	Avg. Annual General Inflation Rate (%)	Avg. Annual Construction Cost Inflation Rate (%)*
1970-79	5.9	7.8
1980-89	4.6	5.7
1990-99	2.5	3.0
2000-08	3.3	5.1
2009-18	2.2**	3.6***

*Based on the Turner Building Cost Index
**Congressional Budget Office projection
***Author's estimate

will cost $2,000,000+ in 14 years (when the need for replacement arrives) with a 5% construction cost inflation rate!

Since annual budgeted funds are not spent until the end of the component's service life, these funds can be placed in interest-bearing investments and, thus, an interest rate factor (IRF) must be incorporated into the calculation. Thus, the annual budgeted funding, x, required to yield the required funds for the replacement or renovation of a component at the end some n years in the future, the RDSL, can be determined by the following relationship:

$$x = y \ (1+CEF)^n / \{(1+IRF) \ [(1+IRF)^n - 1]/n\}$$

Using this equation, the annual budgeted amount required for a roof replacement having a current cost of $100,000 and a remaining service life of eight years is computed as follows:

Variable	Value
y =	$100,000
n =	8 (years)
CEF =	0.09 (9% annual cost escalation)
IRF =	0.05 (5% investment interest rate)

Thus,

$$x = 100{,}000 \ [(1+.09)^{8/}\{1.05 \ [(1.05)^8 -1]/1.05\} = \$19{,}872$$

Budgeting $19,872 for investment at the beginning of each year at a 5% annual interest rate will, at the end of 8 years, yield the $199,256 required to replace the roof. In this example, then, a total investment of only $158,976 is required to offset a $199,256 expenditure—a savings of 20%! With longer investment periods, the savings can be as much as 50-60%.

To simplify computations, the numerator of the relationship above is called the "Future Cost Factor," while the denominator is called the "Future Value Factor." Tables 2-3 and 2-4 at the end of this chapter provide a broad range of interest rates and periods for use and, thus, this computation can be reduced to the following:

x = y[Future Cost Factor (from Table 2-3)/Future Value Factor
(from Table 2-4)]

In the example above, the Future Cost Factor is 1.993, while the Future Value Factor is 10.027 These two values yield the following:

$$x = 100{,}000 \times 1.993/10.027 = \$19{,}872$$

The common alternative to funded depreciation is to do nothing about a component that will ultimately fail until the failure has occurred, or soon will. Under this condition, if funds are not readily available for replacement or renovation, the only alternative is to borrow. *In our example from above, borrowing $199,256 to pay for the roof replacement, with loan terms of 6% annual interest rate for 10 years, would result in a total cost of $265,117 (principal plus loan interest)—over $100,000 more than the funded depreciation option!*

Funded depreciation is always the lowest cost approach to funding adequate replacement/renovation maintenance.

LIFE CYCLE COST ANALYSIS AT
REPLACEMENT/RENOVATION TIME

When building components require replacement or major renovation, the opportunity presents itself to re-think original design decisions—first to make improvements over the original design (e.g., better quality, longer design service life, lower maintenance requirements, etc.) and, second, to select these replacement or renovation components on the basis of lowest life cycle cost rather than simply lowest first cost.

The concept of life-cycle costing is not new, and the methods and applications presented in this text are nothing more than more formal approaches to a process that is already familiar to most individuals: comparative shopping. When you go to buy a new car, you look first at the price. Then, you check the gas mileage rating so that you can estimate operating cost. Next, you read up on the repair history of this model so you will have some idea of the maintenance costs. You look into financing so that the cost of borrowed money is considered. You check into resale values and, thus, estimate the value of the car in future years. And, finally, you decide on how many years you plan to keep the car and thus establish, for you, its economic life. All of this is factored into your estimate of the total owning and operating cost for your new car over the period you plan to keep it.

You have just performed a life cycle cost analysis! The concept is very straightforward and is something with which you are already familiar.

To determine the life cycle cost of each alternative approach to replacing or renovating a building component, specific data about each potential alternative must be collected:

Capital Cost

A fairly detailed cost estimate must be developed based on vendor and contractor input. This estimate must include, as applicable, a number of costs that are often overlooked:

1. Special equipment and/or rigging.
2. Demolition.
3. Additional architectural and/or structural requirements associated with mechanical alternatives
4. Additional mechanical or electrical requirements associated with architectural alternatives.
5. Temporary components, means, etc. to allow occupancy or utilization during the construction period.

Other costs that are included in the capital requirement are design fees, which may increase or decrease as a function of the selected alternative; special consultants' fees; mock-ups; special testing; etc.

All investment costs are computed as of the bid date and are assumed to occur at the beginning of Year 1 of the analysis period (economic life).

Loan Factors

With funds that have accrued under a sinking fund plan, there is no interest expense since no borrowed funds are required to implement the required replacement or renovation. But, a loan, if required, must be amortized over its established term with the defined interest rate applied. Therefore, for these projects, two factors must be known and included in the analysis:

1.   What percentage of the project cost is provided from loan funds? This can range from 0 to 100%.

2.   What is the bond interest rate and term?

Energy Cost

The estimation of energy cost associated with an alternative requires that two quantities be known: (1) the amount of each source of energy expected to be consumed annually for the alternative and (2) the unit cost or rate schedule for each energy source. The second quantity is relatively easy to determine by contacting the utilities serving the site or, for some campus facilities, obtaining the cost for steam, power, chilled water, etc. that may be furnished from a central source.

Economic Life Determination

Table 2-1 provides design service life factors that can be used as the "economic" life of a component.

Maintenance and Replacement Considerations

Estimates of routine preventative maintenance costs and periodic replacement or renovation requirements can be made using the data from this text.

Inflation

As discussed above, construction cost inflation is almost always greater than the general inflation rate and, thus, it is good planning to assume that construction cost inflation will be 30-50% higher than the general inflation rate.

There are numerous spreadsheets and software packages available to help handle the computational aspects of life cycle cost analysis. One of the simplest methods is provided by the spreadsheet analysis tool called "LCCA-1.0" used by the State of North Carolina. The advantage to this tool is that it does not utilize "discount factors" that have the effect of reducing the importance of future operating and maintenance costs relative to the initial construction cost. This tool is available to any user at http://www.nc-sco.com/Guidelines/lcca/TOC_LCCA.htm.

CONSEQUENCES OF
DEFERRING REPLACEMENT/RENOVATION

Deferred maintenance, postponing required maintenance (preventative or replacement/renovation) of building components is the result of deferring funding for maintenance. *The term "deferred maintenance" is really a misnomer—the correct term is "deferred maintenance funding."*

Categories of maintenance may be considered to be deferred if not done when they should have been or when they were scheduled to have been done based RDSLs.

Accumulated deferred maintenance results from two basic failures:

1. Under-funding of preventative maintenance so that minor repair work evolves into more serious maintenance problems, often with expensive "collateral damage."

2. Failure to fund replacement or renovation of building components that have reached the end of their service life. This is, by far, the greater contribution to deferred maintenance.

While the root cause of deferred maintenance is systemic inadequate maintenance funding, the problem is compounded by choices made during austere financial times when even minimum routine maintenance is deferred in order to meet "more pressing" fiscal requirements. **What is true is that management, too often, makes the decision that the value and risks of funding maintenance at its required level is not as important as the value and risks associated with some alternative use for available funds!**

This is short-sighted and, ultimately, more costly approach.

Inevitably, the longer it is possible to avoid replacing or renovating a failed building component, the more it will cost and the greater risk there is that costs will be significantly greater:

- Normal cost escalation due to inflation will occur and future costs will be higher than today's costs.

- There are often costs for "collateral damage" caused by the poor performance of a failed component. For example, a roof that is not replaced when needed will result in an increased incidence of water leaks and these leaks may cause corrosion, wet rot, mold infestation, or other problems that must be repaired with the roof is finally replaced. In an HVAC system, failure to replace an AHU drain pan can also result in water leaks. A fan bearing that is not replaced may score the fan shaft or seize on it, requiring replacement of the shaft and, probably, the fan wheel.

Studies have shown that deferred maintenance costs increase by 2 to 4 times the cost of the repair or replacement when it should have been done, assuming no substantial collateral damage resulting from deferring maintenance. So the question must be posed: if building owners cannot allocate the required funds when needed initially, how will they be able to come up with 2-4 times that funding level within some reasonable time frame in the future? Obviously, they cannot, and too often once maintenance is deferred, that maintenance never re-enters the maintenance program and can only be funded as part a major capital improvement program with new funding, typically borrowed. School systems and governments throughout the United States routinely must issue bonds to fund the replacement or renovation of failed or failing buildings that got that way by deferring maintenance to some future time that, essentially, never came.

So, adequate funding for replacement and renovation is required to eliminate deferred maintenance. To make sure the funds are there when needed, a budgeting plan, utilizing the concept of funded depreciation, is required.

Table 2-3. Replacement/Renovation Future Cost Factors

n	Cost Escalation Rate (%)								
Periods	2	3	4	5	6	7	8	9	10
1	1.020	1.030	1.040	1.050	1.060	1.070	1.080	1.090	1.100
2	1.040	1.061	1.082	1.103	1.124	1.145	1.166	1.188	1.210
3	1.061	1.093	1.125	1.158	1.191	1.225	1.260	1.295	1.331
4	1.082	1.126	1.170	1.216	1.262	1.311	1.360	1.412	1.464
5	1.104	1.159	1.217	1.276	1.338	1.403	1.469	1.539	1.611
6	1.126	1.194	1.265	1.340	1.419	1.501	1.587	1.677	1.772
7	1.149	1.230	1.316	1.407	1.504	1.606	1.714	1.828	1.949
8	1.172	1.267	1.369	1.477	1.594	1.718	1.851	1.993	2.144
9	1.195	1.305	1.423	1.551	1.689	1.838	1.999	2.172	2.358
10	1.219	1.344	1.480	1.629	1.791	1.967	2.159	2.367	2.594
11	1.243	1.384	1.539	1.710	1.898	2.105	2.332	2.580	2.853
12	1.268	1.426	1.601	1.796	2.012	2.252	2.518	2.813	3.138
13	1.294	1.469	1.665	1.886	2.133	2.410	2.720	3.066	3.452
14	1.319	1.513	1.732	1.980	2.261	2.579	2.937	3.342	3.797
15	1.346	1.558	1.801	2.079	2.397	2.759	3.172	3.642	4.177
16	1.373	1.605	1.873	2.183	2.540	2.952	3.426	3.970	4.595
17	1.400	1.653	1.948	2.292	2.693	3.159	3.700	4.328	5.054
18	1.428	1.702	2.026	2.407	2.854	3.380	3.996	4.717	5.560
19	1.457	1.754	2.107	2.527	3.026	3.617	4.316	5.142	6.116
20	1.486	1.806	2.191	2.653	3.207	3.870	4.661	5.604	6.727
21	1.516	1.860	2.279	2.786	3.400	4.141	5.034	6.109	7.400
22	1.546	1.916	2.370	2.925	3.604	4.430	5.437	6.659	8.140
23	1.577	1.974	2.465	3.072	3.820	4.741	5.871	7.258	8.954
24	1.608	2.033	2.563	3.225	4.049	5.072	6.341	7.911	9.850
25	1.641	2.094	2.666	3.386	4.292	5.427	6.848	8.623	10.835
26	1.673	2.157	2.772	3.556	4.549	5.807	7.396	9.399	11.918
27	1.707	2.221	2.883	3.733	4.822	6.214	7.988	10.245	13.110
28	1.741	2.288	2.999	3.920	5.112	6.649	8.627	11.167	14.421
29	1.776	2.357	3.119	4.116	5.418	7.114	9.317	12.172	15.863
30	1.811	2.427	3.243	4.322	5.743	7.612	10.063	13.268	17.449
31	1.848	2.500	3.373	4.538	6.088	8.145	10.868	14.462	19.194
32	1.885	2.575	3.508	4.765	6.453	8.715	11.737	15.763	21.114
33	1.922	2.652	3.648	5.003	6.841	9.325	12.676	17.182	23.225
34	1.961	2.732	3.794	5.253	7.251	9.978	13.690	18.728	25.548
35	2.000	2.814	3.946	5.516	7.686	10.677	14.785	20.414	28.102
36	2.040	2.898	4.104	5.792	8.147	11.424	15.968	22.251	30.913
37	2.081	2.985	4.268	6.081	8.636	12.224	17.246	24.254	34.004
38	2.122	3.075	4.439	6.385	9.154	13.079	18.625	26.437	37.404
39	2.165	3.167	4.616	6.705	9.704	13.995	20.115	28.816	41.145
40	2.208	3.262	4.801	7.040	10.286	14.974	21.725	31.409	45.259

Effective Building Maintenance

Table 2-4. Inflation Rate Comparison

n	Interest Rate (%)					
Periods	5	6	7	8	9	10
1	1.050	1.060	1.070	1.080	1.090	1.100
2	2.153	2.184	2.215	2.246	2.278	2.310
3	3.310	3.375	3.440	3.506	3.573	3.641
4	4.526	4.637	4.751	4.867	4.985	5.105
5	5.802	5.975	6.153	6.336	6.523	6.716
6	7.142	7.394	7.654	7.923	8.200	8.487
7	8.549	8.897	9.260	9.637	10.028	10.436
8	10.027	10.491	10.978	11.488	12.021	12.579
9	11.578	12.181	12.816	13.487	14.193	14.937
10	13.207	13.972	14.784	15.645	16.560	17.531
11	14.917	15.870	16.888	17.977	19.141	20.384
12	16.713	17.882	19.141	20.495	21.953	23.523
13	18.599	20.015	21.550	23.215	25.019	26.975
14	20.579	22.276	24.129	26.152	28.361	30.772
15	22.657	24.673	26.888	29.324	32.003	34.950
16	24.840	27.213	29.840	32.750	35.974	39.545
17	27.132	29.906	32.999	36.450	40.301	44.599
18	29.539	32.760	36.379	40.446	45.018	50.159
19	32.066	35.786	39.995	44.762	50.160	56.275
20	34.719	38.993	43.865	49.423	55.765	63.002
21	37.505	42.392	48.006	54.457	61.873	70.403
22	40.430	45.996	52.436	59.893	68.532	78.543
23	43.502	49.816	57.177	65.765	75.790	87.497
24	46.727	53.865	62.249	72.106	83.701	97.347
25	50.113	58.156	67.676	78.954	92.324	108.182
26	53.669	62.706	73.484	86.351	101.723	120.100
27	57.403	67.528	79.698	94.339	111.968	133.210
28	61.323	72.640	86.347	102.966	123.135	147.631
29	65.439	78.058	93.461	112.283	135.308	163.494
30	69.761	83.802	101.073	122.346	148.575	180.943
31	74.299	89.890	109.218	133.214	163.037	200.138
32	79.064	96.343	117.933	144.951	178.800	221.252
33	84.067	103.184	127.259	157.627	195.982	244.477
34	89.320	110.435	137.237	171.317	214.711	270.024
35	94.836	118.121	147.913	186.102	235.125	298.127
36	100.628	126.268	159.337	202.070	257.376	329.039
37	106.710	134.904	171.561	219.316	281.630	363.043
38	113.095	144.058	184.640	237.941	308.066	400.448
39	119.800	153.762	198.635	258.057	336.882	441.593
40	126.840	164.048	213.610	279.781	368.292	486.852

Chapter 3

Preventative Maintenance

The allocation of adequate resources to maintenance
is required for the maintenance to be effective.

BASIC REQUIREMENTS FOR
GENERAL CONSTRUCTION MAINTENANCE

The requirements for establishing a general construction maintenance program vary widely based on the size and complexity of the facility. However, the following basic elements are common to all facilities:

- Have a working knowledge of general construction materials,
  means, and methods.

- Obtain the original design data ("as built" drawings and specifications). This is the information that defines building components and
  details of construction.

- Obtain O&M manuals provided by contractor(s), including original vendor submittals. These manuals should provide basic maintenance requirements for special building components.

BASIC REQUIREMENTS FOR SERVICE SYSTEMS MAINTENANCE

Plumbing
The requirements for establishing a plumbing maintenance and systems troubleshooting program are relatively simple:

- Understand plumbing basic principles and performance requirements for each plumbing system and its individual elements—
  pumps, piping, heat exchangers, fixtures, brass, etc.

- Obtain the original design data ("as built" drawings and specifications). This is the information that defines plumbing systems and component performance requirements.

- Obtain O&M manuals provided by contractor(s), including original vendor submittals on each piece of equipment (especially pump curves) and fixture. These manuals should provide the basic maintenance requirements for plumbing system components.

- Develop and apply routine maintenance and troubleshooting checklists and procedures, including performance tests. These "preventative maintenance" measures keep the systems and components operating properly, reduce energy and water waste, and maintain their design service life.

- If the plumbing systems are extensive, establish spare parts storage for components that have more frequent failure.

Heating, ventilating, and air-conditioning (HVAC)

The requirements for establishing an HVAC maintenance and systems troubleshooting program are straight forward:

- Understand HVAC basic principles and performance requirements for each HVAC system and its individual elements.

- Obtain the original design data ("as built" drawings and specifications). This is the information that defines HVAC systems and component performance requirements.

- Obtain O&M manuals provided by contractor(s), including original vendor submittals on each piece of equipment (especially fan and pump curves). These manuals should provide the basic maintenance requirements for HVAC system components.

- Obtain "as built" control drawings, including valve and damper schedules, that define the control modes and logic for the HVAC system. Without good documentation of the HVAC system controls, the operation of these systems is based solely on assumptions (which may be wrong) and/or guesses (which may be equally wrong).

- Obtain minimum test instruments, as listed below. Without test instruments, it is impossible to determine what is actually happening in each HVAC system. Without this information, repair or additional maintenance requirements cannot be identified.

Table 3-1. Minimum Required HVAC Maintenance Instruments

Quantity to be Evaluated	Instrument(s) Required
Air temperature and humidity	Electronic (digital) sensor
Water temperature	Pete's plug insertion thermometer Surface contact electronic sensor
Air pressure (ductwork)	Magnahelic gauges: 0-0.5" wg 0-3" wg
Water/steam pressure	Oil-filled gauges: 0-150 ft 0-200 psi
Air flow (ductwork)	Pitot tube (with magnahelic gauges) Anemometer (hot wire or rotating vane)
Air flow (air outlets or inlets)	Flow hood
Water/steam flow	Typically, must be determined by measuring pressure drop across a known component
Volts/amps	Clamp-on amp meter with voltage leads (multi-meter)

- Develop and apply routine maintenance and troubleshooting checklists and procedures, including performance tests. These "preventative maintenance" measures keep the systems and components operating properly, reduce energy waste, and maintain their design service life.

- Establish logs and other long term record keeping procedures. By maintaining data on a regular basis, system changes (and problems) can be identified early. It is particularly important to keep detailed energy consumption records to identify energy loss problems in HVAC systems.

These logs must also include the "normal" condition (or acceptable range of conditions) for each data element recorded so that maintenance personnel can quickly determine when a specific condition indicates a potential problem.

Fire protection

Fire protection system maintenance requirements for commercial and institutional facilities are defined by NFPA Standard 25, *Standard for the Inspection, Testing, and Maintenance of Water-Based Fire Protection Systems*. This standard addresses sprinkler systems, standpipe and hose systems, water supplies, etc. And, most states have specific licensing requirements for the technicians that inspect, test, and maintain fire protections systems.

Any facility with fire protections systems must first determine whether inspection, testing, and maintenance of these systems is to be done with in-house staff or is to be performed by another party on a contract basis—generally, only large facilities find in-house fire protection maintenance to be cost-effective.

Electrical

The following documentation of electrical systems is required to support an adequate electrical systems maintenance program:

- Single-line diagrams of each electrical system. These diagrams should show the electrical circuitry and all major load components and should show all electrical equipment with the following information: voltage, frequency, phase, normal operating position, transformer impedance, available short-circuit current, etc.

- "As-built" design drawings showing all electrical component and loads locations, raceway layout, etc.

- Records of electrical equipment and load component nameplate data.

- Copies of original vendor electrical equipment and component submittals

Once the electrical systems are documented, obtain minimum test instruments, as listed below. Without test instruments, it is impossible to determine what is actually happening in each electrical system. Without this information, repair or additional maintenance requirements cannot be identified.

Table 3-2. Minimum Required HVAC Maintenance Instruments

Quantity to be Evaluated	Instrument(s) Required
High voltage occurrence	Statiscope
Volts, amps, ohms (instantaneous)	Volt meter, volt-ohmmeter, and clamp-on ammeter with multiscale ranges
Volts, amps, ohms, watts, power factor, or volt-amperes (over a period of time)	Recording instruments
Circuiting	Continuity tester
Receptacle circuiting	Receptacle circuit tester
Insulation resistance	Megohmmeter
Ground fault	Ground fault locator

The most common cause of electrical equipment failure is loose connections. Every electrical maintenance program must include routine infrared imaging testing and required tightening of electrical connections. Therefore, insulated tools, gloves, and other protection must be available so that this effort can be done safely while equipment is energized.

For larger systems, an initial short-circuit and coordination study should be performed by a professional engineer to serve as a resource for the maintenance staff is selecting fuse sizes and adjusting over-current protection devices.

All electrical systems maintenance personnel must have a working knowledge of NFPA 70, National Electrical Code; NFPA 70B, Recommended Practice for Electrical Equipment Maintenance; and NFPA 70E, Standard for Electrical Safety in the Workplace.

While NFPA 70, *National Electrical Code,* is applicable primarily to designers, every electrical maintenance staffer should have a basic "working knowledge" of this standard as it requirements also apply to electrical repairs and replacements.

NFPA 70B-2006, *Recommended Practice for Electrical Equipment Maintenance,* includes Chapter 7 on "Personnel Safety" that relies heavily on NFPA 70E and the appropriate OSHA safety-related documents, which should be followed for the development of programs and procedures associated with maintenance activities. Section 70B-7.1.2 states, "Personnel safety should be given prime consideration in establishing maintenance practices, and the safety rules should be instituted and practiced to prevent injury to personnel, both persons who are performing tasks and others who might be exposed to the hazard. The principal personnel danger from electricity is that of shock, electrocution, and/or severe burns from the electrical arc or its effects, which can be similar to that of an explosion."

A key element with regard to personnel safety is first determining what constitutes a qualified person. As noted in Section 70B-7.1, "Maintenance should be performed only by qualified personnel who are trained in safe maintenance practices and the special considerations necessary to maintain electrical equipment. These individuals should be familiar with the requirements for obtaining safe electrical installations." Equally important is "the qualified person should determine if the hazard exposure is limited and restricted against those not qualified for the particular task so a person not qualified for a specific task, even though fully qualified in all other ways, should not be exposed to the hazard of that specific task," as outlined in 70B-7.2.2.1.

Training is another key element in the revised document, as "all employees should be trained in safety-related work practices and required procedures as necessary to provide protection from electrical hazards associated with their respective jobs or task assignments," as noted in 70B-7.3.

Key steps in establishing an electrically safe work condition are identified in 70B-7.4.2, including:

1. Determine possible sources of electrical supply to the specific equipment.

2. After properly interrupting the load current, open the disconnecting device(s) for each source.

3. Apply lockout/tagout devices in accordance with a documented and established policy.

4. Use an adequately rated voltage detector to test each phase conductor/circuit part to verify they are de-energized.

5. Where the possibility of induced voltages or stored electrical energy exists, ground the phase conductors or circuit parts before touching them.

6. Each job task should be analyzed for what particular safety hazards could be encountered. The arc flash hazard is one of the most important of these. Per 70B-7.6, "Switchboards, panelboards, industrial control panels, and motor control centers that are likely to require examination, adjustment, servicing, or maintenance while energized should be field marked to warn qualified persons of potential electric arc flash hazards."

These key elements are addressed in more detail in NFPA 70E-2009, *Standard for Electrical Safety in the Workplace*. And, NFPA 70E is the basis for OSHA's current general industry standard. Referenced in OSHA 29CFR Part 1910, Subpart S, Appendix A, NFPA 70E is considered by OSHA to be the recognized industry practice for electrical safety. In its standard interpretation of the relevance of NFPA 70E, OSHA states, "Industry consensus standards, such as NFPA 70E, can be used by employers as guides to making the assessments and equipment selections required by the standard." Similarly, in OSHA enforcement actions, they can be used as evidence of whether the employer acted reasonably.

NFPA added new training requirements to 79E, as follows: "Employees shall be trained to select an appropriate voltage-detector and shall demonstrate how to use a device to verify the absence of voltage, including interpreting indications provided by the device. The training shall include information that enables the employee to understand all limitations of each specific voltage-detector that might be used."

And, Article 110.7(A) now requires that electrical safety programs be documented.

Nomenclature too has changed. The terms "hot" and "live" have been replaced with the term "energized electrical conductors or circuit parts" throughout. And, to make the standard more specific, the term "working on or near" has been dropped and "limits of approach boundaries" are defined.

PREDICTIVE MAINTENANCE METHODS

The equipment components of building systems can be evaluated on a routine basis using various "predictive" maintenance technologies. The goal of these inspections are twofold: (1) improve estimates of remaining service life and (2) help determine if current preventative maintenance procedures are as effective as they need to be. There are literally dozens of predictive maintenance "tools" available, but generally they fall into the following areas:

Chemical Analysis

Lubricating oil analysis is rotating equipment is a common and effective predictive maintenance tool. Oil samples, taken at routine and regular intervals of equipment operation, are analyzed in order to track the levels of trace chemicals and elements over time in order to determine wear location and characteristics within the component.

Generally, this analysis evaluated the levels of certain elements that can be used as "markers" to define problem conditions within the component. Suspended of dissolved non-oil materials that are indicators of contamination or excess wear include silicon, copper, iron, lead, tin, aluminum, cadmium, molybdenum, nickel, titanium, etc. Contaminants such as acids, dirt or sand, bacteria, fuel, water, or plastics in the oil are indicators of a leaking lubrication system.

And, the condition of the oil itself must be determined. The oil must be evaluated to determine if consistency, viscosity, and corrosion protection or cleaning additives are within normal ranges and do not indicate oil failure.

Vibration Analysis

Every element of rotating equipment components vibrates at a "characteristic" frequency. The combined effect of the these various component frequencies produces a unique "signature" for the piece of equipment.

Using a technique called "broad band analysis," the vibration signature of the equipment item should be determined once the equipment is placed into operation and completed its "break-in." Then, on a routine time interval basis the analysis should be repeated and the new results compared to the previous readings. Changes in the vibration signature means that some component within the equipment has changed it vibration characteristic, implying wear or damage that may mean near term failure and the need for maintenance.

Vibration analysis can be applied to any rotating equipment, such as motors, pumps, and fans within buildings. Water chillers are also prime candidates for vibration analysis.

Infrared Scanning

Infrared, or thermal ("thermographic"), scanning can identify "hot spots" that signify the need for maintenance. Bearings, boilers, heat exchangers, motors, hot piping, and electrical panels and switchgear (including transformers) should be routinely surveyed (every 3-5 years) to identify the locations where the temperatures are too high and maintenance is required.

Ultrasound Scanning

Ultrasound transmitters emit a very high frequency sound wave. Echoes produced by this sound wave reflecting from materials or components can then be received and evaluated to determine the thickness and density of the material being tested. Ultrasonic analysis will also identify voids, cracks, or pits that may exist.

Ultrasound scanning has a wide range of applications in building maintenance:

1. Failing motors, fans, and pumps within the building HVAC and plumbing systems can be quickly identified and traced to bearing problems, drive or coupling wear, etc.

2. Leaks in HVAC systems can often be found using ultrasonic testing long before the leak becomes audible or visible to the maintenance staff.

3. Valve malfunctions and internal oil leaks (bypass) within hydraulic elevators.

4. In high voltage electrical systems, corona discharge problems can be located based on ultrasound produced.

5.   Non-homogeneous materials, such as walls and foundations, can be
      tested to locate cracks or voids.

Bore Scopes and Cameras

Using fiber optics, "bore scopes" can be used to visually inspect the
insides of may components in building systems such as heat exchanger
tubes, piping, etc. For even broader application, ultra-small video camer-
as can be inserted into equipment, wall cavities, piping, ductwork, etc. to
provide visual condition data that can be used to determine maintenance
requirements.

Eddy Current Testing

Eddy-current testing uses electromagnetic induction to detect flaws
in conductive materials. In a standard eddy current testing a circular coil
carrying current is placed proximity to the test specimen. The alternating current in the coil generates changing magnetic field which interacts
with test specimen and generates eddy current. Variations in the phase
and magnitude of these eddy currents can be monitored using a second
"search" coil, or by measuring changes to the current flowing in the primary "excitation"' coil. Variations in the electrical conductivity or magnetic permeability of the test object, or the presence of any flaws, will
cause a change in eddy current flow and a corresponding change in the
phase and amplitude of the measured current.

Eddy-current testing can detect very small cracks in or near the surface of the material, the surfaces need minimal preparation, and physi-
cally complex geometries can be investigated. It is also useful for making
electrical conductivity and coating thickness measurements.

There are several limitations, among them: only conductive materials can be tested, the surface of the material must be accessible, the finish
of the material may cause bad readings, the depth of penetration into the
material is limited, and flaws that lie parallel to the probe may be undetectable.

PREVENTATIVE & PREDICTIVE MAINTENANCE PROCEDURES

As discussed in Chapter 1, preventative maintenance consists of
routine, scheduled maintenance activities and procedures designed to
ensure the facility components meet both performance requirements and

achieve their design service life.

There are negative consequences for not performing adequate preventative maintenance. Obviously, the service life of the component is shortened, increasing the replacement/renovation maintenance costs significantly. And, component reliability is reduced—there are more equipment outages, roof leaks, etc., negatively impacting the facility's mission.

The appendices provide recommended minimum maintenance procedures for facility components, as identified in the following table. Additional manufacturer required and/or recommended maintenance procedures, as documented in the building operations and maintenance manual(s), must also be met. Table 3-3 tabulates the minimum recommended maintenance procedures defined in Appendices A through H:

Table 3-3.
Component Minimum Recommended Preventative Maintenance Procedures

Level 1 Major Group Elements	Level 2 Group Elements	Level 3 Individual Elements	Level 4 Sub-Elements	Preventative Maintenance Procedures
1. SUB-STRUCTURE	1. Foundations	1. Standard Foundations	Wall foundations	1.1.1.0
			Column foundations	1.1.1.0
			Foundation drainage	1.1.1.0
		2. Special Foundations	Pilings/pile caps	1.1.1.0
			Grade beams	1.1.1.0
			Caissons	1.1.1.0
			Overburden	1.1.1.0
	2. Slab on Grade		Standard slab on grade	1.2.0.0
			Structural slab on grade	1.2.0.0
			French drains	1.2.0.0
			Perimeter insulation	1.2.0.0
	3. Basement Walls		Basement wall construction	1.3.0.0
			Wall moisture protection	1.3.0.0
2. SHELL	1. Super-structure	1. Floor Construction	Structure: Wood (Pine, fir, etc.)	2.1.1.0
			Laminated wood (LVL)	2.1.1.0
			Masonry	2.1.1.0
			Structural steel frame	2.1.1.0
			Concrete frame	2.1.1.0
			Floor deck: Wood	2.1.1.0
			Concrete	2.1.1.0
			Floor insulation	2.1.1.0
			Balcony floors: Cast-in-place concrete	2.1.1.1
			Precast concrete	2.1.1.1
			Exterior stairs/fire escapes:	
			Steel, painted or galvanized	2.1.1.0, 2.2.1.11
			Steel, hot-dipped galvanized	2.1.1.0
			Stainless steel	2.1.1.0

Table 3-3. (*Continued*)
Component Minimum Recommended Preventative Maintenance Procedures

Level 1 Major Group Elements	Level 2 Group Elements	Level 3 Individual Elements	Level 4 Sub-Elements	Preventative Maintenance Procedures
		2. Roof Construction	Flat deck: Plywood, OSB	2.1.2.0
			Metal	2.1.2.0
			Metal w/concrete	2.1.2.0
			Roof insulation	2.1.2.0
			Pitched deck: Plywood, OSB	2.1.2.0
			Metal	2.1.2.0
			Metal w/concrete	2.1.2.0
			Roof insulation	2.1.2.0
			Facia/Soffits/Coping:	2.1.2.0
			Wood	2.1.2.0
			Aluminum (coated)	2.1.2.0
			Steel, painted, galvanized	2.1.2.0
			Steel, stainless	2.1.2.0
			Vinyl	2.1.2.0
			Fiberglass panels	2.1.2.0
			Natural stone	2.1.2.0
			Cast stone	2.1.2.0
			Precast concrete	2.1.2.0
			Cement asbestos	2.1.2.0
			Flashing: Galv. metal	2.1.2.0
			Stainless steel	2.1.2.0
			Copper	2.1.2.0
			Flat roof edging: Alum. parapet cap	2.1.2.0
			Alum. gravel stop	2.1.2.0
			Expansion joint: Prefabricated	2.1.2.0
			Roof joint w/alum.	2.1.2.0
			Ridge vent: Alum	2.1.2.0
			Molded polyethylene	2.1.20
			Canopies: Aluminum	2.1.2.1
			Canvas awnings	2.1.2.2
			Flat roof drains	2.1.2.0
			Flat roof scuppers	2.1.2.0
			Gutters/downspouts: Aluminum	2.1.2.0, 2.1.2.3
			Galv. steel	2.1.2.0, 2.1.2.3
			Stainless steel	2.1.2.0, 2.1.2.3
			Copper	2.1.2.0, 2.1.2.3
			PVC	2.1.2.0, 2.1.2.3
	2. Exterior Closure	1. Exterior Walls	Exterior walls: Precast concrete	2.2.1.0
			Brick veneer	2.2.1.1
			CMU	2.2.1.7
			EFIS (over masonry)	2.2.1.2
			Alum/vinyl siding	2.2.1.3
			Wood siding	2.2.1.4
			Cement board siding	2.2.1.5
			Cement stucco	2.2.1.6
			Metal building panels	2.2.1.12
			Caulking (average)	2.2.1.8
			Flashing: Sheet metal	2.2.1.9
			Lintels: Painted steel	2.2.1.10, 2.2.1.11
			Galvanized steel	None
			Stainless steel	None
			Concrete	2.2.1.0
			Exterior paint on wood or metals	2.2.1.11

Table 3-3. (*Continued*)

Component Minimum Recommended Preventative Maintenance Procedures

Level 1 Major Group Elements	Level 2 Group Elements	Level 3 Individual Elements	Level 4 Sub-Elements	Preventative Maintenance Procedures
		2. Exterior Windows	Windows: Steel windows	2.2.2.0
			Vinyl casement windows	2.2.2.1
			Vinyl sash windows	2.2.2.1
			Aluminum windows	2.2.2.2
			Wood windows	2.2.2.3
			Curtainwalls and Storefronts:	
			Steel	2.2.2.4
			Aluminum (Anodized)	2.2.2.4
			Window blinds	2.2.2.5
		3. Exterior Doors	Exterior doors: Hollow metal	2.2.2.6
			Solid core, wood	2.2.2.6
			Aluminum	2.2.2.6
			Fiberglass	2.2.2.6
			Door hardware: Automatic openers	2.2.2.7
			Closer	2.2.2.6
			Lockset	2.2.2.6
			Panic device	2.2.2.6
			Door stop	2.2.2.6
			Overhead doors: Wood	2.2.2.8
			Steel	2.2.2.8
			Aluminum	2.2.2.8
			Fiberglass	2.2.2.8
			Electric operator	2.2.2.8
			Coil doors: Steel	2.2.2.8
			Electric operator	2.2.2.8
			Air curtains (Cafeteria/kitchen)	2.2.2.9
	3. Roofing	1. Roof Coverings	Pitched roofing: Asphalt shingles	2.3.1.0, 2.3.1.1, 2.3.1.2.1
			Asphalt rolled	
			Wood shingles	2.3.1.0, 2.3.1.1, 2.3.1.2.2
			Slate tiles	
			Metal	2.3.1.0, 2.3.1.1, 2.3.1.2.3
			Built-up	
			Flat roofing: EPDM	2.3.1.0, 2.3.1.1, 2.3.1.2.4
			Other single membranes	
			BUR, 3-ply	2.3.1.0, 2.3.1.1, 2.3.1.2.5
			BUR, 5-ply	2.3.1.0, 2.3.1.1, 2.3.1.2.6
				2.3.1.0, 2.3.1.1, 2.3.1.3.1
				2.3.1.0, 2.3.1.1, 2.3.1.3.1
				2.3.1.0, 2.3.1.1, 2.3.1.3.2
				2.3.1.0, 2.3.1.1, 2.3.1.3.2
		2. Roof Openings	Skylights	2.1.2.0
			Roof hatches	2.1.2.0
			Roof ventilators	2.1.2.0
			Equipment curbs/penetrations	2.1.2.0

Table 3-3. (*Continued*)
Component Minimum Recommended Preventative Maintenance Procedures

Level 1 Major Group Elements	Level 2 Group Elements	Level 3 Individual Elements	Level 4 Sub-Elements	Preventative Maintenance Procedures
3. INTERIORS	1. Interior Construction	1. Partitions	Fixed partitions: Metal studs/GWB	3.1.1.0
			Wood studs/GWB	3.1.1.0
			Plaster on lathe	3.1.1.1
			CMU	3.1.1.2
			Moveable partitions	3.1.1.3
			Toilet partitions: Metal	3.1.1.4
			Fiberglass	3.1.1.4
			Stone	3.1.1.4
			Interior glass and storefronts	3.1.1.5
		2. Ceilings	Acoustical ceiling: Lay-in	3.1.2.0
			Concealed spline	3.1.2.0
			Hard ceiling: GWB on steel runners	3.1.2.0
			Plaster on lathe	3.1.2.0
		3. Interior Doors	Doors: Wood, hollow	3.1.3.0
			Wood, solid	3.1.3.0
			Hollow metal	3.1.3.0
			Fiberglass	3.1.3.0
			Door hardware: Automatic openers	2.2.2.7
			Closer	3.1.3.0
			Lockset	3.1.3.0
			Panic device	3.1.3.0
			Door stop	3.1.3.0
		4. Specialties	Storage shelving/cabinets/lockers	3.1.4.0
			Ornamental metals and handrails	3.1.4.0
			Signage	3.1.4.0
			Closet specialties	3.1.4.0
			White boards	3.1.4.0
			AV equipment brackets/stands	3.1.4.0
			Lavatory counters	3.1.4.0
		5. Fire-stopping	Wall/floor firestop systems	3.1.5.0
	2. Staircases	1. Stair Construction	Stair construction: Steel/concrete	2.1.2.0
			Stair handrails and balustrades: Steel	2.1.2.0
		2. Stair Finishes	Tread and landing finishes	See Floor Finishes
			Handrail and balustrade finishes	3.3.3.1
	3. Interior Finishes	1. Wall Finishes	Painting: Corridors/high traffic areas	3.3.3.1
			Classrooms	3.3.3.1
			Offices, conference	3.3.3.1
			Storage rooms, MER, etc.	3.3.3.1
			Vinyl wall covering	3.3.3.2
			Ceramic tile: Mortar laid	3.3.3.3
			Adhesive backing	3.3.3.3
			Caulking	3.3.3.4

Table 3-3. (*Continued*)
Component Minimum Recommended Preventative Maintenance Procedures

Level 1 Major Group Elements	Level 2 Group Elements	Level 3 Individual Elements	Level 4 Sub-Elements	Preventative Maintenance Procedures
		2. Floor Finishes	Floor coverings: Carpeting, carpet tile	3.3.3.5
			Vinyl tile	3.3.3.5
			Vinyl sheet	3.3.3.5
			VCT	3.3.3.5
			Terrazzo	3.3.3.5
			Wood, solid, T&G	3.3.3.5
			Wood, laminated	3.3.3.5
			Epoxy	3.3.3.5
			Painting (Conc/wood)	3.3.3.1
			Brick/terra cotta tile	3.3.3.3
			Ceramic tile	3.3.3.3
			Bases, curbs, and trim	See Floor Coverings
4. SERVICE SYSTEMS	1. Conveying Systems	1. Elevators, escalators, dumbwaiters, etc.		4.1.1.0
		2. Wheelchair lifts		4.1.1.1
	2. Plumbing	1. Plumbing Fixtures and Equipment	Water closets	4.2.1.0
			Urinals	4.2.1.0
			Lavatories	4.2.1.0
			Sinks	4.2.1.0
			Showers	4.2.1.0
			Faucets/brass: Commercial	4.2.1.0
			Drinking fountains and coolers	4.2.1.0
			Kitchen fixtures	4.2.1.1
			Dishwashers	4.2.1.2
			Grease/art traps	4.2.1.3
			Water heaters: Gas/oil/elec. ≤120 gal	4.2.1.4
			Gas/oil/elec. >120 gal	4.2.1.4
			Steam/storage	4.2.1.4
			Steam/instantaneous	4.2.1.4
		2. Domestic Water Distribution	Copper water piping/valves	4.2.2.0
			Plastic water piping/valves	4.2.2.1
			Pipe insulation	4.3.19.0
			Pumps, line-mounted	4.3.16.0
			Backflow preventer (RPP/RPZ type)	4.2.2.2
			Legionella control	4.2.2.3
		3. Sanitary Waste	Waste piping	4.2.2.1
			Vent piping	4.2.2.1
			Floor drains	4.2.3.0
			Pipe insulation	4.3.19.0
			Pumps/lift station	4.3.16.1
		4. Rain Water Drainage	Piping (Interior)	4.2.2.1
			Piping (Exterior)	4.2.2.1
			Pipe insulation	4.3.19.0
		5. Special Plumbing Systems	Gas piping	4.2.4.0
			Fuel oil piping	4.2.2.1
			Acid waste systems	4.2.4.1
			Pool piping and equipment	4.2.4.2
	3. HVAC	1. Air- conditioning unit, packaged or split system	Window unit	4.3.1.0
			Packaged or split system, <10 tons	4.3.1.1
			≥10 tons	4.3.1.1
			PTAC or PTHP	4.3.1.0
			Ductless split system	4.3.1.1
			Computer room unit	4.3.1.2

Table 3-3. (*Continued*)
Component Minimum Recommended Preventative Maintenance Procedures

Level 1 Major Group Elements	Level 2 Group Elements	Level 3 Individual Elements	Level 4 Sub-Elements	Preventative Maintenance Procedures
		2. Heat pumps	Packaged or split system, < 7.5 tons	4.3.1.1
			≥10 tons	4.3.1.1
			Water source units	4.3.2.0
		3. Air-handling units	Indoor	4.3.3.0, 4.3.3.3, 4.3.3.4, 4.3.3.5
			Outdoor, rooftop	4.3.3.0, 4.3.3.1, 4.3.3.3, 4.3.3.4, 4.3.3.5
			Fan coil units	4.3.3.2, 4.3.3.3, 4.3.3.4
		4. Boilers	Steel firetube, steam	4.3.4.0
			Cast iron, steam or hot water	4.3.4.1
			Electric, hot water	4.3.4.2
			Pressure burner, gas, oil, or dual fuel	4.3.4.3
			Atmospheric burner, gas	4.3.4.4
		5. Furnaces	Gas- or oil-fired	4.3.5.0
		6. Unit heaters	Gas- or oil-fired	4.3.6.0
			Hot water or steam	4.3.6.1
			Electric	4.3.6.2
		7. Radiant heaters	Gas-fired infrared heaters	4.3.7.0
		8. Air Terminals	CV, VAV boxes with or without reheat	4.3.8.0
			Fan-powered boxes	4.3.8.1
		9. Ductwork	Duct work	4.3.9.0
			Diffusers, grilles, and registers	4.3.9.1
			Humidifiers	4.3.9.2
		10. Fans (other than part of air-handling units)	Centrifugal	4.3.10.0
			Axial	4.3.10.0
			Propeller	4.3.10.1
			Exhaust, roof- or sidewall-mounted	4.3.10.1
		11. Coils (other than part of air-handling units)	DX, water, or steam	4.3.3.3, 4.3.3.3, 4.3.3.4, 4.3.3.5
			Electric	4.3.11.0
		12. Heat Exchangers	Shell-and-tube, steam or HW	4.3.12.0
			Plate and frame	4.3.12.1
		13. Chillers	Air-cooled, electric-drive	4.3.13.0
			Water-cooled, electric drive	4.3.13.1
			Absorption, 2-stage direct-fired	4.3.13.2
		14. Cooling towers	Galvanized	4.3.14.0
			Galvanized w/stainless basin/deck	4.3.14.0
			Stainless steel	4.3.14.0

Table 3-3. (*Continued*)
Component Minimum Recommended Preventative Maintenance Procedures

Level 1 Major Group Elements	Level 2 Group Elements	Level 3 Individual Elements	Level 4 Sub-Elements	Preventative Maintenance Procedures
			Ceramic	4.3.14.0
		15. Condensers	Air-cooled	4.3.15.0
			Evaporative	4.3.15.1
		16. Pumps	Base-mounted	4.3.16.0
			Line-mounted	4.3.16.0
			Sump pump	4.3.16.1
			Condensate pump, indoor, above grade	4.3.16.2
			Condensate pump, outdoor, below grade	4.3.16.2
		17. Controls	Pneumatic	4.3.17.0
			Electric	4.3.17.1
			DDC	4.3.17.2
		18. Piping, valves, and specialties	Water piping and valves	4.3.18.0
			Steam piping and valves	4.3.18.1
			Condensate piping and valves	4.3.18.1
			Water specialties: Expansion tanks Vibration connectors Strainers Air separators Pres. relief valves BFP (RPP/RPZ) Steam specialties: PRV's Pres. relief valves	4.3.18.3.1 4.3.18.3.2 4.3.18.3.3 4.3.18.3.4 4.3.18.3.5 4.2.2.2 4.3.18.4.1 4.3.18.4.2
		19. Insulation, pipe/duct	Fiberglass	4.3.19.0
			Calcium silicate	4.3.19.0
			Celluar glass	4.3.19.0
		20. HVAC electric motors		4.3.20.0
		21. Water Treatment Programs	Closed circuit water (HW, CHW)	4.3.21.0
			Condenser water	4.3.21.1
			Steam/condensate	4.3.21.2
		22. Bearing Lubrication		4.3.22.0
	4. Fire Protection	1. Sprinkler Systems	Sprinkler water supply Piping Air source for dry pipe systems Fire pump	4.4.1.0 4.4.1.0 4.4.1.0 4.4.1.0
		2. Standpipe & Hose Systems	Standpipe water supply Piping Fire hose/cabinets	4.4.1.0 4.4.1.0 4.4.1.0
		3. Fire Protection Specialties	Fire extinguishers/cabinets Kitchen hood fire suppression system Special fire protection systems	4.4.3.0 4.4.3.1 4.4.3.1
	5.Electrical	1. Electrical Service & Distribution	High voltage service and distribution Low voltage service and distribution	4.5.1.0 4.5.1.1

Table 3-3. (*Continued*)
Component Minimum Recommended Preventative Maintenance Procedures

Level 1 Major Group Elements	Level 2 Group Elements	Level 3 Individual Elements	Level 4 Sub-Elements	Preventative Maintenance Procedures
		2. Lighting & Branch Wiring	Branch circuit wiring	4.5.1.1
			Panels/breakers	4.5.1.1
			Wiring devices	4.5.1.1
			Light fixtures	4.5.2.1
		3. Communication, Fire Alarm, and Security Systems	Public address system	4.5.3.0
			Intercommunication/paging system	4.5.3.1
			Call systems	4.5.3.2
			Clock systems	4.5.3.3
			Fire alarm system	4.5.3.4
			Security monitoring and alarm systems	4.5.3.5
			CCTV systems	4.5.3.6
		4. Special Electrical Systems	Grounding	4.5.4.0
			Emergency lighting	4.5.4.1
			Emergency power systems	4.5.4.2
			Static uninterruptible power systems	4.5.4.3
			Rotary uninterruptible power systems	4.5.4.4
5. EQUIPMENT & FURNISHINGS	1. Equipment	1. Commercial Equipment	Office equipment/computers	None
		2. Institutional Equipment	Library equipment	None
			Theater and stage equipment	None
			AV equipment	None
			Laboratory equipment	None
			Art equipment	None
		3. Vehicular Equipment	Parking control/gates	5.1.3.0
			Loading dock levelers, etc.	5.1.3.1
		4. Other Equipment	Food service coolers/freezers	5.1.4.0
			Kitchen hood	5.1.4.1
	2. Furnishings	1. Fixed Furnishings	Fixed casework	None
			Blinds and other window treatment	None
			Seating	None
			Gymnasium flooring/removable wood	None
6 SPECIAL CONSTRUCTION	1. Special Structures		Pre-engineered metal buildings	6.1.0.0
			Portable classrooms	6.1.0.1
	2. Special Facilities		Aquatic facilities	6.2.0.0
			Fuel oil storage tanks	6.2.0.1
7. BUILDING SITEWORK	1. Site Improvements	1. Roadways	Bases/sub-bases	See "Paving"
			Paving and surfacing: Gravel/shellrock	7.1.1.0
			Concrete	7.1.1.1
			Asphalt	7.1.1.2
			Curbs, gutters, and drains	7.1.1.3
			Pavement lines and markings	7.1.1.4
			Signage	7.1.1.5
			Bridges	7.1.1.6
		2. Parking Lots	Bases/sub-bases	See "Paving"
			Paving and surfacing: Gravel/shellrock	7.1.1.0
			Concrete	7.1.1.1
			Asphalt	7.1.1.2
			Curbs, gutters, and drains	7.1.1.3
			Pavement lines and markings	7.1.1.4
			Signage	7.1.1.5

Table 3-3. (*Continued*)
Component Minimum Recommended Preventative Maintenance Procedures

Level 1 Major Group Elements	Level 2 Group Elements	Level 3 Individual Elements	Level 4 Sub-Elements	Preventative Maintenance Procedures
		3. Pedestrian Paving	Concrete paving	7.1.1.1
			Asphalt paving	7.1.1.2
			Steps (concrete)	7.1.1.1
			Pavement lines and markings	7.1.1.3
			Signage	7.1.1.4
			Bridges	7.1.1.5
		4. Site Development	Fencing/gates	7.1.4.0
			Retaining walls	7.1.4.1
			Signage	7.1.1.4
			Site furnishings	7.1.4.2
			Fountains, pools, water features	7.1.4.3
			Flagpoles	7.1.4.4
			Playgrounds and play equipment	7.1.4.5
		5. Landscaping	Erosion/runoff control features	7.1.5.0
			Irrigation systems	7.1.5.1
	2. Site Civil / Mechanical Utilities	1. Water Supply & Distribution Systems	Well water	7.2.1.0
			Well pump: Submersible	7.2.1.1
			Above ground	7.2.1.1
			Piping and valves	7.2.1.1
			Pressure tank: Steel	7.2.1.1
			Fiberglass	7.2.1.1
			Backflow preventer (RPP/RPZ)	4.2.2.2
			Water softener	7.2.1.1
		2. Sanitary Sewer Systems	Piping	7.2.2.0
			Manholes and cleanouts	7.2.2.0
			Septic tank: Concrete	7.2.2.1
			Septic field	7.2.2.1
			Lift station	7.2.2.2
			Package waste treatment system	7.2.2.3
		3. Storm Sewer Systems	Piping	7.2.2.4
			Manholes and cleanouts	7.2.2.4
			Headwalls/catch basins	7.2.2.4
			Ditches and culverts	7.1.4.3
			Retention ponds	7.1.4.3
			Lift station	7.2.2.2
		4. Heating Distribution	Piping and valves: Hot water	4.3.18.0
			Steam	4.3.18.1
			Condensate	4.3.18.1
			Manholes	7.2.4.0
			Exposed insulation w/weather jacket	4.3.19.0
		5. Cooling Distribution	Piping and valves: CDW, CHW	4.3.18.0
			Exposed insulation w/weather jacket	4.3.19.0
		6. Fuel Distribution	Gas piping: Above grade, painted	4.2.4.0
			Below grade, protected	None
			Below grade, plastic	None
			Oil piping: Above grade, painted	4.2.2.1
			Below grade, protected	None
			Below grade, plastic	None
	3. Site Electrical Utilities	1. Exterior Lighting	Fixtures, poles, and stanchions	7.3.1.0
			Raceway and wiring	7.3.1.0
			Lighting controls	7.3.1.0
		2. Exterior Communications & Security	Public address system	4.5.3.0
			Intercommunication/paging system	4.5.3.1
			Security monitoring and alarm systems	4.5.3.5
			CCTV systems	4.5.3.6

Each of these procedures, to the maximum extent possible, has been developed in a format that calls for the maintenance technician to perform three basic duties:

1. *Investigate*: Requires the technician to thoroughly inspect the component and defines conditions to look for.

2. *Evaluate*: For each condition identified, there are one or more probable causes that should be evaluated.

3. *Maintain*: Perform specific maintenance preventative procedures plus additional procedures indicated by the condition(s) identified.

Chapter 4

Special Maintenance Considerations

SUSTAINABLE MAINTENANCE vs. SUSTAINABLE BUILDINGS

Buildings have a significant direct and indirect impact on the environment. As illustrated by Figure 4-1, buildings not only use resources such as energy and raw materials, they also generate waste and potentially harmful atmospheric emissions.

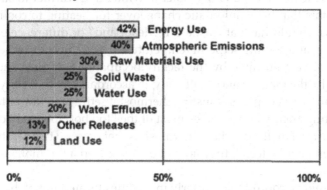

Figure 4-1. Environmental Impact of Buidings (Source: USDOE)

A "sustainable" approach to building operation and maintenance supports an increased commitment to environmental stewardship and conservation, and attempts to balance of costs and environmental, societal, and human benefits, all while still meeting the mission and function of the intended facility or infrastructure.

Energy Efficiency

As energy costs continue to rise and the availability of some energy sources becomes ever more problematic, it makes sense for every facility to implement measures to eliminate excessive energy consumption. To do this, the facility maintenance staff can typically implement improvements in three areas:

HVAC Operating Setpoints and Schedules

Establishing on-off (occupied-unoccupied) schedules with individual temperature setpoints can reduce energy consumption by heating, ventilating, and air-conditioning (HVAC) equipment by 10-25%.

In larger facilities that have a facility-wide direct digital control system, this system should be used to schedule operation of every piece of HVAC equipment and establish temperature (and, as required, humidity) setpoints for each occupancy period. But, even in simple facilities (branch banks, dentist offices, retail stores, etc.), replacing older existing thermostats with newer programmable, seven-day thermostats (a $40-$150 investment) is very cost effective.

In order to have good indoor air quality, temperature setpoints should in the range of 72-77°F (cooler in winter and warmer in summer). Thermostats that have automatic changeover for heating to cooling and vice versa, should have at least a 5°F "deadband" or difference between the heating and cooling setpoint temperatures. And, thermostat setpoints should not be adjustable by the facility occupants. Setpoints should be changed by the maintenance staff, only, in accordance with facility policy. This eliminates occupants adjusting thermostat setting too low or too high and wasting energy and resolves most of the "too hot/too cold" conflicts that can arise. Finally, purchase thermostats that have no temperature display; often people decide they are hot or cold based a thermometer reading.

Humidity control, particularly in facilities located in hot, humid climates is necessary to prevent the development of mold infestation. Modifying existing systems to incorporate humidity control, adding humidity sensors or humidistats, and establishing humidity setpoints may be required. To minimize the energy expended for humidity control, humidity setpoints should be in the range of 30% RH (winter low limit) to 60% RH (summer high limit).

In addition to space temperature and humidity setpoints, there are other setpoints in HVAC systems that are critical to energy consumption:

- For constant or variable air volume air handling systems that provide supply air at a constant temperature, that temperature setpoint should be as high as possible while maintaining comfort conditions and preventing humidity conditions in the facility exceeding 60% RH.

• To minimize chiller/cooling tower energy consumption, maintain the chilled water supply temperature setpoint as high as possible and the condenser water temperature as low as possible. Chiller efficiency improves 3-4% for each 1°F increase in chilled leaving water temperature or each 1°F reduction in entering condenser water temperature.

• The supply water temperature for hydronic heating systems should be set as low as possible. One way to do this is via a "reset schedule" that adjusts the supply temperature inversely proportional to the outdoor air temperature.

Preventative Maintenance Program

Chapter 3 and the Appendices of this text define the parameters and procedures for an effective facility preventative maintenance program. Implementation of this program, tailored to each individual facility, will result in the facility performing as intended when designed and not deteriorating over time.

As part of these maintenance procedures, pay attention to the presence of heating devices (portable heaters, etc.), cooling devices (electric fans, etc.), and lighting devices (lamps, torchieres, etc.) that have been brought into the building by the occupants. These devices consume energy and, in some cases, represent safety hazards. But, more importantly, they are strong indicators that the occupants are dissatisfied with the performance of the facility systems. Pay attention! Investigate each problem and, if at all possible, correct them so that inefficient (and potentially unsafe) occupant-provided devices can be removed.

Low Cost/No Cost Building Improvements

The following are improvements that can be implemented as part of the preventative maintenance program or planned replacement/renovation of facility components. While some of these measures may have a significant initial cost, the simple payback for them can be expected to be between 1-2 years, making it possible to finance their implementation through only 1 or 2 budget cycles.

- *Reduce Outdoor Air Infiltration*: Preventative maintenance procedures recommended in this text address the routine caulking of doors, windows, etc. to maintain an airtight building. But, as planned replacement/renovation is implemented, additional improvements should be considered:
  — Select replacement operable windows that have minimal infiltration ratings. Where possible, replace operable windows with fixed glazing.
  — When exterior doors are replaced, consider the use of revolving doors in high traffic areas and adding adequately sized vestibules for entrances with even moderate traffic.
  — Add air curtains to roll-up doors in warehouses, garages, etc.

- *Reduce Thermal Losses*: As planned replacement/renovation is implemented, additional improvements should be considered:
  — When replacing windows, select glass with the lowest possible U-factor to reduce temperature-driven heat gains and losses. In DOE Climate Zones 1, 2, and 3, consider the use of Low-E glazing or even reflective glass to reduce solar heat gains.
  — When reroofing a low slope roof, consider adding additional insulation to the roof. Since the roof is being replaced, increasing the roof insulation by 1-3" can be very cost effective.
  — When reroofing, consider the roof color. In cold climates, a darker colored roof will absorb more solar radiation that can be used to help offset heating needs. In warm climates, a lighter colored roof will do a better job of reflecting solar radiation, reducing cooling requirements.

- *Convert from Constant Volume Airflow to Variable Volume Airflow and Reduce/Eliminate Reheat*: Peak heating and cooling is required only for very short periods each winter and summer. For constant volume airflow systems, though, the amount of air required for these peak loads is circulated at all times the system is on, requiring significant fan energy consumption. To reduce this "transport" energy requirement, some types of secondary HVAC systems can be converted to variable air volume (VAV) operation.

  Prime candidates for conversion to VAV are constant volume terminal reheat systems and dual duct systems. For both of these systems, the conversion to VAV, in addition to reducing fan energy, also significantly reduces the amount of reheat.

The first step in converting the either system is to modify terminal units for variable airflow. This can be done by either replacing existing terminal units with new VAV units (the preferred method) or adding a new VAV air valve upstream of the existing terminal unit. If the second option is selected, it will be necessary to remove the constant volume regulator in the exiting terminal unit which, for older mechanical type regulators, will reduce the system static pressure requirement by as much as 1" wg.

The terminal unit controls, then, must be modified to allow 50-70% airflow reduction before the heating/reheating coil is energized.

At the AHU, a variable frequency drive (VFD) is required for each supply and return fan, controlled on the basis of supply duct static pressure.

The savings from this conversion can be as much as a 50% reduction in air transport, heating, and cooling energy consumption.

- *Reduce Distribution Pressure Losses*: Excess pressure losses in air duct and water piping systems increase transport energy consumption. The resistance to flow in a duct or piping system is the sum of the resistances in the index path or circuit, i.e., the path or circuit with the largest resistance to flow. By locating and improving the pressure losses in the index path or circuit, fan or pump energy can be reduced. To determine the index path or circuit and identify high losses that can be addressed, the following procedures can be used:
  — Duct systems:
    (1) Using a magnahelic pressure gauge with flexible tubes and simple ¼" copper tube static pressure probes, measure the pressure drop across the primary AHU components (coils, filters, fan, etc.) Compare these values against the "new" values in the original balancing report or equipment submittals. If the measured values are more than 10% higher, maintenance of filters, coils, fans, etc. is required.
    (2) Starting at the AHU, measure the duct static pressure upstream and downstream of each change of direction (elbow) or branch connection (tee) to determine if there are excess pressure drops. Check on each side of every fire and/or smoke damper to determine if the fire damper is partially (or completely!) closed.

(3) Determine the entering pressure to each terminal unit. If this pressure is more than 15% higher than the sum of terminal unit pressure loss and the downstream low pressure ductwork losses, the fan is producing excess pressure that can be reduced by changing the static pressure controller setpoint (for VAV systems) or slowing the fan (for constant volume systems).

(4) Finally, check the static pressure drop across each terminal unit and compare this value to the design values in the original balancing report or equipment submittal. If this value is higher, the unit may need cleaning or repair.

— Water piping systems:

(1) *Using a single, high quality pressure gauge* that is moved from point to point, determine the pressure drop for each heat exchanger in the system. Compare these values to the design values in the original balancing report or equipment submittal. If any value is more than 15% higher, the heat exchanger may need cleaning or repair.

(2) Check the pressure drop across strainers and clean if required.

(3) Sometimes, the pressure drop through the index circuit will significantly exceed the pressure drop through the other circuits in the system. If this is the case, determine if the index circuit can be re-piped to reduce losses. Another option is to use a booster pump in this one circuit and shave the impeller of the main pump(s) to reduce the available head to the remaining circuits and, thus, save energy.

• *Add/Remove an Airside Economizer Cycle*: Airside economizers, which provide "free" cooling using cold outdoor air, should not be applied to dual duct or multizone secondary HVAC systems. If this is the case, energy savings will result by eliminating the economizer. (Note, however, that if the chiller must now run in the winter, modifications to the chiller and its cooling tower may be required and a waterside economizer may be more cost effective.)

For other types of air systems that require cooling during the winter, though, an airside economizer can reduce cooling energy consumption by 10-30%. For systems without return fans, implementing an airside economizer cycle requires the addition of a return fan or a relief fan(s). Engineering investigation and analysis is required to determine the most cost-effective approach.

•   *Use Water-Cooled or Evaporative Condensing*: The vast majority of direct expansion (DX) cooling systems utilize outdoor air as a "heat sink" for condenser heat rejection. However, by using evaporative cooling, there is a significant improvement in the cooling energy requirement as shown by Table 4-1.

Table 4-1. Comparison of Condensing Energy Impact

Type Of Condensing	Typical Condensing Temperature(°F)	Typical Compressor kW/Ton
Air-Cooled	120-130	0.9-1.4
Evaporative	94-98	0.6-0.8

Thus, evaporative condensing can reduce cooling energy consumption by about 40% over air-cooled condensing. Existing air-cooled equipment can be retrofitted with evaporative "pre-cooling" pads or, as older air-cooled equipment reaches the end of its service live, water-cooled replacements will typically be very cost effective.

•   *Convert Chilled Water System to Variable Flow*: Primary-secondary or variable flow primary piping configurations significantly reduce chilled water pumping energy and improve the overall loading and efficiency of multi-chiller systems.

•   *Install Waterside Economizer Cycle*: The energy consumption by the waterside economizer for the pumps and cooling tower is only about 20-25% of the energy consumption by mechanical cooling. Thus, for fancoil unit systems or other air systems that cannot utilize an airside economizer cycle, winter cooling energy requirements can be reduced by 75-80%.

•   *Improve Burner Efficiency*: Every burner, not matter what size, should be maintained for maximum firing efficiency. However, for larger systems (1,500,000 Btu/h input or greater), improvements to the burner can be made that will increase the overall firing efficiency.

— Preheat Combustion Air: Preheating air entering the burner to support combustion will reduce the cooling effect when it enters the combustion chamber and improve boiler efficiency. In most boiler rooms, air is heated incidentally by hot boiler and pipe radiation losses and rises to collect below the ceiling or roof. This air can be easily ducted down to the burner. Boiler efficiency increases approximately 2% for every 100°F increase in combustion air temperature, up to a maximum temperature of approximately 300-350°F.

— Change Oil-Fired Boiler Burners from Steam Atomizing to Air Atomizing: Compressed air can be used for atomizing fuel oil in lieu of steam, eliminating a parasitic boiler load that be as much as 3-5% of the boiler capacity.

— Add Boiler Trim Control: Boiler efficiency can be improved by incorporating an excess air trim loop into the boiler controls. It is easy to monitor excess air as oxygen not used for combustion is heated and discharged with the exhaust gases. A stack gas oxygen analyzer can be installed to continuously monitor excess air and adjust the boiler fuel-to-air ratio for optimum efficiency. A carbon monoxide trim loop, used in conjunction with the oxygen analyzer, assures that incomplete combustion cannot occur due to deficient air supply.

Table 4-2 illustrates the effect of excess air control on overall combustion efficiency for natural gas firing:

Table 4-2. Excess Air (%) vs. Combustion Efficiency

Excess (%)		Combustion Efficiency (%) at Flue Gas Temperature Less Combustion Air Temperature (F)				
Air	Oxygen	200	300	400	500	600
9.5	2.0	85.4	83.1	80.8	78.4	76.0
15.0	3.0	85.2	82.8	80.4	77.9	75.4
28.1	5.0	84.7	82.1	79.5	76.7	74.0
81.6	10.0	82.8	79.3	75.6	71.9	68.2

— Replace Burner and/or Boiler with Higher Efficiency Unit: Even if combustion efficiency is maximized, the overall boiler efficiency of smaller cast iron and steel boilers will be lower due to poorer heat transfer efficiency in the boiler sections. Thus, when these boilers reach the end of their service life, alternatives should be considered. Generally, steel boilers are more efficient than cast iron boilers. For boilers up to about 1,500,000 Btu/h capacity, the use of a condensing boiler, providing overall boiler efficiency as high as 96%, should be considered.

For large firetube boilers, the overall boiler efficiency is dictated by the number of "passes" that the flue gas makes through the boiler. A 2-pass boiler has a maximum overall efficiency of only about 80%, while a 3-pass boiler has an efficiency of 83-84% and a 4-pass boiler can achieve 85-86%.

When a boiler must be replaced, it is always cost-effective to replace it with the highest efficiency unit available.

• **Isolate Off-Line Boilers**: It is important to prevent boiler cycling to maintain overall efficiency. For most facilities, multiple smaller boilers will result in more efficient operation than a single larger boiler.

With multiple boilers, when heating loads are low, the imposed load may be met by one boiler while the other(s) remain idling on standby. Idling boilers, however, consume energy to meet standby losses, which can be increased by continuous induced flow of air through them and out the stack. To reduce these losses, each boiler should be equipped with an automatically controlled stack damper. Thus, instead of idling, the off-line boilers can be shut down and their stack dampers closed to eliminate losses.

• **Add Stack Gas Economizer**: In even a well-tuned boiler system, typical stack gas exit temperature will be 350-450°F, depending on the type of fuel. To prevent condensation in the stack, the flue gas temperature cannot be reduced to below about 250°F for natural gas and 325°F for oil, but that limitation still allows anywhere from 100 to 125°F of heat recovery from the flue gases. A heat exchanger can be installed in the stack leaving the boiler and this wasted heat can be used to preheat make-up water or feedwater or to heat combustion air.

- *Improve Hot Piping Insulation*: Table 4-3 summarizes the recommended minimum pipe insulation thickness for steam and hot water piping systems. If existing steam or hot water piping has little or no insulation, it is very cost-effective to insulate the piping to meet these recommendations.

Table 4-3. Recommended Hot Piping Insulation Thickness

Existing Insulation Thickness (in.)	Additional Insulation Thickness (in.) at Fluid Temperature (F)		
	200 or less	201-350	351+
½ or less	1	1	2
1	1	1	2
1½	—	1	1½
2	—	—	1

- *Minimize Exhaust and Balance Ventilation Air*: To minimize energy waste in any facility, first minimize the amount of exhaust air, while maintaining health and safety standards. The vast majority of plants and facilities in the United States exhaust volume's of air far in excess of those required to maintain the health and safety of workers and great amounts of energy and money are being wasted.

   To eliminate this waste, each facility should do the following:
— Measure/calculate the total air being exhausted.
— Reduce that volume to a minimum to meet code or operational requirements and provide a satisfactory indoor air quality.

   The first step is to inventory all exhaust outlets from a facility. This inventory must include exhaust outlets from boilers and other equipment and all local and general ventilation outlets. The quantity of exhaust air or gas is always measured in cubic feet per minute (cfm), i.e., the air velocity (fpm) multiplied by the area of the exhaust opening (sf). Air velocity may be measured by the following methods:

— By means of a pitot tube and manometer: These instruments are used to measure the velocity pressure caused by a flow of gas in a duct. It is calibrated in inches of water, which can easily be converted to velocities. To be accurate a number of readings using a "pitot transverse" must be made. See a ventilation manual or test manual for details.

— By means of an air velocity meter: This is an instrument with a moving vane or rotor that may be held in the air-stream and which will indicate the air velocity on a scale calibrated in feet per minute. While not as accurate as using a pitot tube and manometer, it is easier to use.

Ventilation has significant impact on a facility's total energy consumption. Each cubic foot of air brought into the facility must be heated or cooled and, in some cases, humidified and/or dehumidified. Therefore, to reduce energy use, first reduce exhaust airflow as much as possible, but maintain a slight positive pressure to retard infiltration-caused heat losses/gains and water vapor intrusion.

Establish a ventilation operation schedule so exhaust system operates only when it is needed.

• *Add Exhaust Air Heat Recovery*: When more than 10,000 CFM of exhaust air is involved, and the facility configuration permits, consider installation of heat recovery devices.

• *Add Boiler Blowdown Heat Recovery*: Any boiler with continuous surface blowdown exceeding 5% of the steam generation rate is a good candidate for blowdown waste heat recovery. The recovered heat is normally used to preheat feedwater.

• *Install occupancy sensors for lighting control*: Infrared-based occupancy sensors can be installed to switch lights off when they are not needed. Applied to toilets, conference rooms, storage rooms, individual offices, etc., these devices can result in 25-75% lighting energy savings in these areas.

When installing occupancy sensors, the following guidelines apply:
— Adjust sensors to avoid "false-offs."
— Make sure sensor location does not result in "non-human" sources of motion that can trigger sensors.
— Set sensors to fail in the "on" position in dark areas (e.g., areas with no windows).
— Add routine sensor inspection and testing to the facility preventative maintenance program.

- *Replace inefficient lighting sources*: Existing lighting sources should be replaced with more efficient types of lighting based on the following:
  - Replace incandescent lamps with compact fluorescent lamps (CFLs). CFLs provide a 75% reduction in lighting energy with a life that is 3-5 times longer.
  - Replace older fluorescent T-12 lamps and magnetic ballasts with T-8 (or even T-5) lamps with electronic ballasts. (This process typically requires that the entire fixture be replaced and implementation can be part of planned renovation/replacement over several years.)

- *Implement lighting operating schedules*: The amount of energy consumed by lighting systems (kWh) is dependent on the type and level of lights (kW installed) and its hours of use each day. To reduce hours of use, lighting schedules based on the facility work or use schedule should be implemented, typically on a area-by-area basis.

  The least cost approach to implementing this measure is simply to schedule housekeeping rounds as early as possible in the afternoon and evening and assign the housekeeping staff with responsibility for cutting off lights as they finish cleaning each area. Thus, when the facility occupants return the next morning, they can turn lights on as needed.

  A more expensive approach is to install lighting contactors at each lighting panel and connect these contactors to controllers (or, if available, a facility-wide direct digital control system) to turn the lights on and off in accordance with a pre-set schedule.

Indoor Water Consumption

Since 1993, all new facilities have been required to use "low-flow" plumbing fixtures—water closets and urinals, lavatory and sink faucets, and shower heads. These fixtures reduce water use by 50-78% over pre-1993 fixtures. Older facilities, when replacing older fixtures that have reached the end of their service life (or when renovating), must replace them with the current low-flow type. However, no matter which type of fixture is currently installed, indoor water consumption can be reduced with the following measures that can be incorporated into the maintenance program.

Find and Repair Leaks

In Appendix D, recommended plumbing system maintenance procedures incorporate specific requirements for inspecting and testing water systems for leaks. When a leak is detected, it must be repaired *immediately*.

Enhance Water Efficiency

• Reduce water usage at lavatories by replacing standard aerators with 0.5-gpm flow limiting aerators that still service hand washing needs adequately, but reduce water consumption by 30-75%.

• Replace existing showerheads with 2.0-gpm flow limiting heads, reducing water consumption by 20% or more.

• Replace existing hand-operated faucets with infrared sensor activated faucets or with faucets that use spring-loaded devices to automatically turn the faucets off after 10-15 seconds of use.

• Gravity or tank type flush water closets, referred to as "high efficiency toilets" are now available that require 1.3 gallons per flush or less, saving about 20% in water consumption. Thus, replacing code-compliant water closets with high efficiency fixtures should be evaluated.

High efficiency urinals reduce water consumption from 1.6 gpf to 0.25-0.5 gpf. To achieve this water efficiency, these urinals typically require automatically controlled flush valves and special fixture design.

• Older flush valve water closets can be retro-fitted with dual flush mechanisms. When only liquid waste must be flushed, the flush handle is moved in one direction and the water closet consumes only about 0.8 gallon of water. When solid wastes must be flushed, the flush handle is moved in the opposite direction and 1.6 gallons of water is used.

A study conducted in 2003 by California authorities indicates that the potential savings in water consumption by utilizing dual flush water closets is significant, as shown in Table 4-4.

Table 4-4. Dual Flush Water Consumption

Type of Building	Ratio of Short to Long Flushes	Net Water Consumption (GPF)
Office Building	1.7 to 1.0	1.10
Restaurant	1.3 to 1.0	1.15
Residential (Multi-Family)	4.0 to 1.0	0.96

Thus, water savings over plumbing code requirements range from 28% to 40%, depending on the type of building occupancy.

- Waterless urinals save almost 95% water consumption, using water only as part of routine maintenance wash-down. These devices have been in use since the 1970s in Europe, but only arrived in the United States in mid-1990s. Waterless urinals have been found to be more hygienic than conventional ones since the aerosol effect by which bacteria can be spread is eliminated.

  Typically, these urinals utilize a trap insert filled with a proprietary sealant liquid instead of water. The lighter-than-water sealant floats on top of the urine collected in the U-bend, preventing odors from being released into the air. Other designs do not use a cartridge; instead, using an outlet system that traps odors.

  If waterless urinals are considered for retrofit projects, careful evaluation of the existing waste piping is required. Carbon steel, stainless or galvanized, will react with urine, which is slightly acid (pH 4-6), resulting in accelerated corrosion. Therefore, old waste piping systems that may contain galvanized steel piping should be upgraded for waterless urinals. Copper, cast iron, and PVC piping do not have corrosion issues with urine.

  Maintenance of waterless urinals is somewhat greater than for conventional urinals since routine cleaning and deodorizing (usually once per week) is required and the trap cartridges have to be replaced periodically. Trap life depends on the number of uses, typically around 7,000, not on the length of time it has been in use, which makes it somewhat more difficult to establish a replacement schedule. But, considering flush valve maintenance that is eliminated with waterless urinals, there is little net increase in overall maintenance requirements.

- Electronic controls for plumbing fixtures usually function by transmitting a continuous beam of infrared (IR) light. With faucet controls, when a user interrupts this IR beam, a solenoid is activated, turning on the water flow. Dual-beam IR sensors or multi-spectrum sensors are generally recommended because they perform better for users with dark skin.

  Depending on the faucet, a 10-second hand wash (required minimum by the Americans with Disabilities Act) typical of an electronic unit will consume as little as 1/3 quarts of water.. Choose the lowest-flow faucet valves available, typically 0.5 gpm.

  At sports facilities, where urinals experience heavy use over a relatively short period of time, the entire restroom can be set up and treated as if it were a single fixture. Traffic can be detected and the urinals flushed periodically based on traffic rather than per person. This can significantly reduce water use.

  Electronic controls can also be used for other purposes in restrooms. Sensor-operated hand dryers are very hygienic and save energy (compared with conventional electric hand dryers) by automatically shutting off when the user steps away.

- Commercial garbage disposals grind solid wastes into small particles for disposal into the sewer system. The ground garbage passes into a mixing chamber where it blends with water for disposal. In larger systems, a scraping and pre-flushing system may precede grinding and carry the materials to the garbage disposal. Some larger systems use a conveyor instead of a scraper to transport waste to the disposal.

  Typical water-consumption rates for various garbage disposals and disposals combined with scrapers or conveyor equipment range from 6-10 gpm. Thus, consider eliminating garbage disposals to reduce both water use and maintenance. One option is to utilize garbage strainers in lieu of garbage disposals. A strainer-type waste collector passes a recirculating stream of water over food waste held in a basket. This reduces waste volume as much as 40 percent by washing soluble materials and small particles into the sewer. The water use for strainers is about 2 gpm, much less than the 5 to 8 gpm requirement of garbage disposals. Strainers can use wastewater from the dishwasher, eliminating added water consumption.

  When a disposal must be used, installation of flow regulators end excess flow due to high water pressure and timers with automatic shut-off limit disposal over-operation. A solenoid valve can also be used to control water flow to the disposal.

- Recirculating hot water systems, typical for most larger non-residential applications, are the source of energy loss (both heat loss in the piping and pumping energy required for water circulation), but typically reduce water waste. To reduce the energy burden imposed by the recirculating systems, pumps should be turned off during unoccupied periods and low flow restrictors installed on each hot water outlet.

- For both cooling towers and boilers, the amount of blowdown for deposition control can be minimized by using automatic water treatment systems. For automatic blowdown control, the blowdown requirement is determined by measuring the boiler water electrical "conductance," a measure of the amount of conductive solids in the water, since pure water has zero conductance. The recommended boiler water conductance level for HVAC boilers operating at less that 350 psig is 3500 mega-mho/cm or less, where a "mho" is the conductive equivalent to an "ohm" of resistance. For cooling towers, the water conductance setpoint must be determined based on the water chemistry.

- For cooling towers and evaporative condensers, the term *cycles of*

**Table 4-5. Blowdown Cycles of Concentration vs.
Make-Up Water Requirements**

| | GPM/Ton | | | % |
Cycles	Evaporation	Blowdown	Make-up	Make-up
2	0.0300	0.0300	0.0600	100
3	0.0300	0.0150	0.0450	75
4	0.0300	0.0100	0.0400	67
5	0.0300	0.0075	0.0375	63
6	0.0300	0.0060	0.0360	60
10	0.0300	0.0033	0.0333	55
15	0.0300	0.0023	0.0323	54
20	0.0300	0.0015	0.0315	53

concentration defines the ratio of the desired concentration of dissolved solids in the condenser water to the concentration of dissolved solids in the make-up water. Table 4-5 summarizes the water flows associated with a cooling tower for various cycles of concentration.

It is clear that the amount of make-up water is reduced significantly as the number of cycles is increased from 2 to 6. However, there is only a further 5% reduction as the cycles is increased from 6 to 10, and only a further 2% reduction as cycles is increased to 20. Therefore, to minimize water consumption, cycles of concentration should be maintained at no lower than 10 and deposition inhibitors added as necessary.

• Steam and condensate leaks due to piping leaks and/or trap failures, which require make-up water consumption, can be addressed by preventative maintenance procedures. However, condensate recovery measures may be required to capture condensate that, for whatever reason, is purposely wasted. This waste falls into three categories: (a) condensate lost through distribution line traps, (b) condensate lost at process equipment (industrial equipment, sterilizers, laundry equipment, kitchen equipment, etc.), and (c) condensate lost via boiler blowdown:

Outdoor Water Consumption:
Automatic irrigation systems are the prime culprit in excess outdoor water consumption. Every one of these systems results in excessive water consumption, increased runoff, and increased pollution. (And, no amount of tinkering with their controls will improve their poor environmental impact.) However, the need for these systems can be significantly reduced or even eliminated very easily, as follows:

• Revamp landscape plantings to minimize, or even eliminate, the need for irrigation. In every climate zone, there are landscaping plants, both "native" and those that have climate adapted, that are far more "drought resistant" than others. There are numerous national and state publications available (most over the Internet) that can provide guidance for replacing existing plants with those that require far less water and can thrive with only natural rainfall.

• Decorative lawns surrounding facilities may be pretty, but if irrigated they are water "hogs." First, replace as much lawn area as possible with "natural" areas or areas landscaped with drought-resistant plants. Next, for those lawns that must remain, select grasses that require little or no watering beyond natural rainfall. (The typical high cost of maintenance for lawns should automatically make them a target for reduction!)

The recovery and recycling of wasted water to offset irrigation water consumption is also an option, as follows:

• Normal operation of cooling coils produces condensate water that typically drains to the sewer. But, condensate is clean water that can be captured and reused for non-potable water applications. Typical applications include cooling tower make-up, flushing fixtures, and landscape irrigation.
   Considerations for condensate recovery systems include:
   — Condensate recovery works by gravity flow. A drainage system from each air handling unit to a central connection point, a holding tank or cistern, is required. A pump then supplies make-up water from the holding tank or cistern to the cooling towers.
   — Collected condensate water is at temperatures between 50 and 60°F.
   — To use condensate as cooling tower make-up, a 3-way valve in the line feeding make-up water to the cooling towers is required to allow the system to draw from reclaimed condensate or service water as needed for level control. Normally, the cooling towers need more make-up water than can be recovered from the condensate, in which case the system uses supplemental domestic water.

   The amount of condensate produced by cooling will range from 0.005 to 0.0167 gpm/ton, based on the amount of outdoor air and the climatic conditions that exist. Thus, a 100 ton commercial HVAC system, operating for 2000 equivalent full load hours annually, can produce as much as 70,000 gallons of clean condensate that can be captured and used in lieu of potable water sources.
   One caveat regarding cooling condensate recovery: during cooling coil cleaning operations, all water from the drain pan must be drained to the sanitary sewer. This may require bypass piping to

a floor drain or the use of a pump and collection tank during the cleaning process.

- A rainfall harvesting system consists of the following basic components:
  — Catchment surface, typically the roof of the building.
  — Collection system...the gutters and downspouts that collect and transfer the rainwater to a storage tank.
  — Leaf screens, first-flush diverters, and/or roof washers that remove dust and debris from the initial catchment runoff. First-flush diverters are designed to divert or waste a portion of the initial rainfall to eliminate contaminants that were on the catchment surface when the rainfall started.
  — Storage tanks, called "cisterns."
  — Filtration, to make the water potable, if required by the application. Given the local regulatory requirements, either cartridge filtration or RO may be required. (No filtration is required if the water is not used for human consumption.)

In theory, about 0.62 gallons/sf of catchment area can be collected per inch of rainfall. However, considering first flush losses, evaporation, splash, etc., a value of 0.50 gallons/sf of catchment area per inch of rainfall is typically used for system sizing. The catchment area is based on the "horizontal projection" of the building roof and is independent of roof pitch. To ensure a year-round water supply, the catchment area and storage capacity must be sized to meet the water demand through the longest expected span of continuous dry days.

The capital cost of rainwater harvesting systems is highly dependent on the type of catchment, conveyance, and storage tank materials used.

The primary reference for the design of rainwater harvesting systems is The Texas Manual on Rainwater Harvesting, available as a free download at http://www.twdb.state.tx.us/publications/reports/RainwaterHarvestingManual_3rdedition.pdf.

Stormwater Runoff

Surface water runoff occurs when it rains and a percentage of the rainfall drains into the local watershed streams and rivers. In "green-fields" or natural, undeveloped land, the surfaces are considered to be near 0% impervious and the majority of rainwater is retained by natural vegetation long enough to be absorbed into the ground or released very slowly. Typically, about 50% of the rainfall is absorbed, about 40% retained long enough to be evaporated, and only about 10% runs off into streams and rivers.

However, with development, a significant portion of natural vegetation is typically replaced by impervious surfaces. Impervious surfaces are mainly constructed surfaces—rooftops, sidewalks, roads, and parking lots—covered by impenetrable materials such as asphalt, concrete, brick, and stone. These materials seal surfaces, repel water, and prevent precipitation and meltwater from infiltrating soils. Soils compacted by urban development (even including most lawns) are also highly impervious.

The science clearly shows that there is a negative impact on watershed water quality as a result of this increased runoff. Surface pollution due to human activity—organic waste, fertilizers and other landscape chemicals, vehicle VOCs from oil and gasoline spills, etc.—are carried by the runoff into the local watershed.

To reduce the negative impact of runoff, every facility should consider the following:

- As discussed above, natural areas are far more pervious than other planted areas, especially lawns—they retain water and reduce runoff. Lawns not only are more impervious, increasing the rate and quantity of stormwater runoff, but they typically are fertilized and treated with other chemicals that become part of the runoff, resulting in watershed pollution. Thus, replacing grassed areas with natural planting areas will reduce both the quantity of stormwater runoff and improve the quality of what runoff that does occur.

  The construction of "rain gardens" is becoming popular to create artificial wetlands that filter runoff and slow its rate of release.

- Rainwater or air-conditioning condensate that is captured and used for plant irrigation reduces the need for purchased water. Also, harvested rainwater reduces site runoff since the use of the water is spread out over a much longer time period.

- Improve stormwater retention: Runoff from developed sites is inevitable. However, retaining stormwater and releasing it over a longer period reduces its negative impacts. First, contaminants that are carried away by the "first flush," usually considered to be the first 1-1/2" to 2" of rainfall, are trapped in retention ponds and structures and can be periodically removed. Second, all remaining rainfall, rather than being released immediately can be held and released over several days, reducing the damage to streams.

Sustainable Building Risks and Maintenance Requirements

The main design objectives for "sustainability" in buildings are to avoid resource depletion (energy, water, and raw materials); prevent environmental degradation caused by facilities and infrastructure throughout their life cycle; and create facilities that are livable, comfortable, safe, and productive. However, the design and construction of sustainable facilities may involve the use of new, untested materials or the use of conventional materials in a new way. Since it is unrealistic to expect designers and contractors to have a complete understanding of the performance characteristics and limitations of new products, it is almost a "given" that sustainable design can create maintenance problems, ranging from shorter service life to moisture problems and mold infestation. Areas of concern include the following:

- Natural building materials may not have the needed service life for the application. For example, bamboo flooring has a service life of 7-12 years in residential applications, which is quite satisfactory, but only 3-4 years in a high traffic commercial applications, which imposes a significant maintenance requirement.

- Recycled materials and materials with high bio-based content may have more moisture issues and potential mold problems.

- Application failure will result in more than maintenance problems. For example, hay bale construction requires that wall materials be protected by a perfect water barrier since the more conventional layered approach is not possible—and there is no such thing as a perfect water barrier!

Maintenance programs and procedures must be developed to anticipate these potential problems as much as possible, but these programs and procedures must also include routine testing and performance evaluation so that any problems that do occur are detected and corrected as early as possible.

Sustainable facilities may have more complex systems and maintenance requirements, such as the following:

- *Heating, ventilating, and air-conditioning systems*: It is essentially impossible for packaged, direct-expansion air-cooled HVAC systems to meet even the most basic energy efficiency goal for sustainable design, typically a 30% improvement over the requirements of ASHRAE Standard 90.1-2007. Thus, water-cooled systems will usually be applied. Thus, for some owners, maintenance experience and skill sets may be inadequate, at least initially.

  HVAC system controls in sustainable facilities tend to be more complex and, consequently, require a greater degree of expertise on the part of the maintenance staff to ensure their functionality.

  The over-emphasis by some designers on improved indoor air quality as a part of sustainable design can result in maintenance moisture-related problems. The inappropriate use of excess ventilation and/or natural ventilation (including operable windows), can result in inadequate building humidity control for facilities in hot, humid climates. While these types of design problems may require retro-fitting to correct, it certainly a responsibility of the maintenance staff to provide routine inspection and evaluation to identify problems as early as possible.

- *Daylighting*: The level and amount of cleaning for daylighting will increase. Daylighting results in increased glass area that must be cleaned more frequently that simple vision glass. Likewise, light shelves, reflective louvers, etc. require routine cleaning in order to maintain their reflective properties. Leaves, snow, bird droppings, etc., in addition to normal dust and dirt, reduce their effectiveness if not removed quickly.

  Some daylighting devices such as louvers or blinds have electromechanical control systems that must be maintained in accordance with the manufacturer's requirements.

  And, with the application of daylighting, sensors are required to dim/brighten the artificial lighting systems in response to the daylighting available at any given moment. To achieve satisfactory lighting levels and to meet energy performance goals, these systems must be routinely maintained.

- *Photovoltaic (PV) Systems*: PV panel performance is highly dependent on its ability to remain clean. The cleaning frequency for a PV system will depend upon the following:
  — Urban locations or locations in dry, dusty climates will increase the frequency of cleaning. Coastal areas will experience salt spray that also must be removed frequently.
  — Rain acts as a natural cleaning agent, so locations in rainy climates tend to have reduced maintenance cleaning requirements.
  Photovoltaic systems have numerous electrical components that require specific maintenance, including batteries, inverters, and controls.

- *Wind Power Systems*: Wind turbine maintenance is specialized and is typically beyond the capabilities of most facility maintenance staffs. Therefore, most of these systems are maintained under outside contract. As the turbine contains more moving parts, hence it requires considerable maintenance. During the early years, maintenance cost is typically between 1.5% and 3% of the turbine cost, but this cost increases as the turbine get older.
  The electrical components of the system, including, batteries, inverters, and controls, also require routine maintenance.

- *Combined Heat and Power Systems (CHP):* CHP plant maintenance is very complex and is typically beyond the capabilities of all but the largest and most sophisticated maintenance staffs. Therefore, most of these systems are maintained under outside contract.
  Such on-site service and maintenance can be provided by the original equipment manufacturer or supplier or by a specialist third-party organization. Maintenance contracts require that a monthly fee be paid by the plant owner; that is, in effect, a continuous full protection plan for engine and generator and provide for fast (typically eight-hour) response times by service personnel in repairing failed components.
  Service and maintenance agreements typically cover only CHP equipment rather than the whole plant. Most on-site power generation plant owners contract out service and maintenance simply because most don't have the in-house expertise necessary to carry out routine maintenance tasks on reciprocating engine gensets, heat

exchangers, boilers, etc., let alone major repair or replacement operations. Gas turbines and large generators are definitely the province of the professional. In such cases, owners often contract to buy maintenance services from the OEM or specialist third party contractor for several years at least.

ANSI/ASHRAE/USGBC/IES Standard 189.1-2009, *Standard for the Design of High-Performance, Green Buildings Except Low-Rise Residential Buildings*, is the first code-intended commercial green building standard in the United States. It has been published by the American Society of Heating, Refrigerating and Air-Conditioning Engineers (ASHRAE), in conjunction with the Illuminating Engineering Society of North America (IES) and the U.S. Green Building Council (USGBC) and has been approved by the American National Standards Institute (ANSI).

This standard, published on January 22, 2010 provides a long-needed consensus standard for those who strive to design, build, and operate sustainable buildings. It covers key topic areas including site sustainability, potable water use efficiency (both indoor and outdoor), energy efficiency, indoor environmental quality. and the building's impact on the atmosphere, materials and resources. Like other ASHRAE Standards, it is written in enforceable code language so that it may be referenced or adopted by code authorities.

Section 10 of the new standard requires that a maintenance plan addressing mechanical, electrical, plumbing, and fire protection systems be developed. The maintenance measures defined in Chapters 1, 3 and 4 of this text provide significant information that can be easily incorporated into the maintenance plan required by the new standard for sustainable building design, construction, and operation.

Section 10 of the new standard also requires a service life plan be developed to address structural, building envelope, and hardscape materials repair/replacement during the service life of the building. Planning for the replacement or major renovation of building components at the end of their service live is addressed in detail in Chapter 2 of this text, which can be used to meet the requirements of the new standard.

Both the maintenance plan and the service live plan are required to be developed during the design of a sustainable building. And, during design, Chapter 5 of this text can serve as a guideline for designing the building for improved "maintainability." This is critical in the design of sustainable buildings simply because such building typically have more complex building systems; use recycled, bio-based, and/or regional materials in the building construction; and/or use conventional materials in new ways—all of which may have unanticipated consequences and special maintenance needs.

INDOOR AIR QUALITY

Trying to define indoor air quality (IAQ) is somewhat akin to the Supreme Court defining pornography—we know what it isn't, even if we don't know exactly what it is. The American Society of Heating, Refrigerating, and Air-Conditioning Engineers (ASHRAE), in their Standard 62.1, defines "acceptable indoor air quality" as follows: "air in which there are no known contaminants at harmful concentrations as determined by cognizant authorities and with which a substantial majority (80% or more) of the people exposed do not express dissatisfaction." The problem with this definition is simply who wants to live with the 20% that *do* express dissatisfaction! (Sounds like a bunch of academics taking advice from lawyers.)

In broad terms, "good" IAQ results from the following:

- Temperature and humidity control by the building HVAC systems within the ranges of 72-77°F and 30-60% RH.

- Adequate dilution ventilation with outdoor air to maintain occupant effluents (CO_2 and odor) to acceptable levels.

- Adequate lighting.

- Low noise and vibration levels.

- Ongoing monitoring and maintenance to avoid introducing pollutants or contaminants into the indoor air from either indoor sources or outdoor sources.

When one or more of these conditions does not exist, there is potential for IAQ problems.

Poor IAQ, real or perceived, can be the result of many factors:

- *Energy cost reduction measures*: As energy costs have risen and the emphasis on energy cost savings have assumed importance with building owners and managers, HVAC systems have been modified to reduce ventilation airflows, to eliminate the use of reheat for humidity control, to be started as late as possible and shutdown as soon as possible, to broaden temperature control ranges, etc.

- *Building design and construction*: Many buildings are designed with better air barriers that reduce infiltration airflows and the resulting secondary ventilation; "least cost" design and/or poor construction has cut corners resulting in increased water intrusion, poor lighting design, and noise or vibration problems, along with HVAC systems that cannot satisfy both temperature and humidity control needs, particularly in hot, humid climates; and the use of more and more synthetic or manufactured building materials that have the potential of releasing pollutants and contaminants as they age.

- *Building use changes*: Buildings change in response to changing business or occupant needs, or simply need to be updated as they age, and renovation of existing buildings is common. This renovation means that new building products and new building systems that can regularly introduce (or exacerbate existing) IAQ problems.

People factors also enter into the IAQ equation. Building occupants are regularly exposed to news coverage on indoor environmental issues, including asbestos, PCB's, lead paint, mold, etc. This awareness, as one author states, "has raised the expectations of occupants regarding the quality of the air in their personal space." It is these expectations, and any failure to meet them by the building owner, designer, contractor, and/or a manger, that can lead to IAQ litigation.

There are real health risks from poor indoor air quality, though these risks are not nearly as severe or widespread as reported in the media. Pollutants can be introduced from the outdoors due to poor air cleaning, including pollen, dust, fungal spores, vehicle exhaust, and biological or chemical pollutants from industrial, research, or medical sites. Pollutants introduced by indoor sources include formaldehyde and volatile organic compounds (VOCs) "off-gassed" by building materials (wall and floor finishes, furniture, etc.); VOCs from cleaning supplies; sewer odor; body odor; ozone produced by copy machines and laser printers; mold resulting from water leaks, water intrusion into the building, and/or poor building humidity control; etc.

Because of varying sensitivity among people, one individual may react to a particular IAQ problem while surrounding occupants have no ill effects. (Symptoms that are limited to a single person can also occur when only one workstation receives a higher percentage of the pollutant dose.) In other cases, complaints may be widespread or not. A single indoor air pollutant or problem can trigger different reactions in different people and some may not be affected at all. Information about the types of symptoms can sometimes lead directly to solutions, however, symptom information is more likely to be useful only for identifying the timing and conditions under which problems occur.

IAQ problems often produce nonspecific symptoms rather than clearly defined illnesses. And, analysis of air samples often fails to detect high concentrations of specific contaminants.

Sick building syndrome (SBS) is a term applying to IAQ problems with nonspecific symptoms or cause(s) that affect a high percentage (5-20%) of occupants. Building-related illness (BRI) is a term referring to illness brought on by exposure to the building air, where symptoms of diagnosable illness are identified (e.g., certain allergies or infections) and can be directly attributed to environmental agents in the air. Legionnaire's disease and hypersensitivity pneumonitis are examples of BRI that can have serious, even life-threatening consequences.

A small percentage of the population may be sensitive to a number of chemicals in indoor air, each of which may occur at very low concentrations. The existence of this condition, known as multiple chemical sensitivity (MCS), is a matter of considerable controversy. MCS is not currently recognized by the major medical organizations, but medical opinion is divided. The applicability of access for the disabled and worker's compensation regulations to people who believe they are chemically sensitive may become concerns for facility managers.

Sometimes several building occupants experience rare or serious health problems (e.g., cancer, miscarriages, Lou Gehrig's disease) over a relatively short time period. These clusters of health problems are occasionally blamed on indoor air quality factors and can produce tremendous anxiety among building occupants. State or local Health Departments can provide advice and assistance if clusters are suspected. They may be able to help answer key questions such as whether the apparent cluster is actually unusual and whether the underlying cause could be related to IAQ.

Temperature and Humidity

The first step toward maintaining acceptable IAQ is to control temperature and humidity to acceptable levels in each occupied space; if building occupants are too hot or too cold, they are "uncomfortable" and may become more sensitive to indoor air quality. Therefore, it is important that indoor temperatures be maintained with a comfort range. Based on ASHRAE, building HVAC controls should be set to maintain indoor temperature conditions at 75°F ± 2°F.

Humidity also has direct effect on the indoor human health parameters, as shown in the Figure 4-2.

Thus, while most people are "comfortable" with humidity as low as 20-30% RH, the optimum low limit on humidity is 40% RH, though a 30% RH is acceptable unless maintained for long periods (northern climates).

Lighting

Poor lighting can be an irritant that exacerbates other IAQ problems and can produce its own health risk in the form of headaches and eye

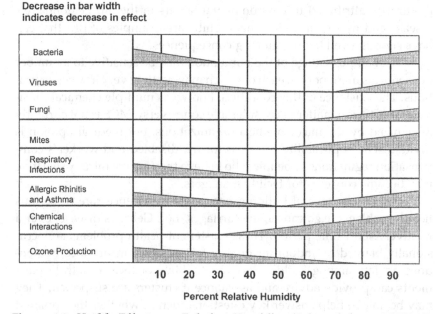

Figure 4-2. Health Effects vs. Relative Humidity (Adapted from ASHRAE Transactions, 1984)

strain. Lighting design is beyond the scope of this text, but careful attention to the brightness, glare, and quality of light is required for good lighting.

The Illuminating Engineering Society of North America (IESNA) established three "groups" and seven categories based on descriptions of various visual tasks, with recommended levels of light for each task category.

Group 1: Orientation and Simple Visual Tasks

Visual performance is largely unimportant for this level of task. These tasks are found in public spaces where reading and visual inspection are only occasionally performed. Parking lots are normally lit to 5 footcandles.

Higher levels are recommended for tasks where visual performance is occasionally important.

Group 2: Common Visual Tasks

Visual performance is important. These tasks are found in commercial, industrial and residential applications. A desktop workspace is normally lit to 30 to 50 footcandles. Recommended illuminance levels differ because of the characteristics of the visual task being illuminated. Higher levels are recommended for visual tasks with critical elements of low contrast or small size.

Group 3: Special Visual Tasks

Visual performance is of critical importance. These tasks are very specialized, including those with very small or very low contrast critical elements. Recommended illuminance levels should be achieved with supplementary task lighting. Higher recommended levels are often achieved by moving the light source closer to the task.

The recommended lighting level for each group and task category is defined by Table 4-7.

The maintenance staff, using a light meter and regular light fixture maintenance should be able to respond to and correct lighting deficiencies in a facility. If not, a lighting engineer can be consulted.

Noise and Vibration

Vibration from HVAC equipment, piping, and ductwork can be a source of undesired noise and can contribute to perceived poor IAQ.

Table 4-7. Task Specific Recommended Lighting Levels

Category	Space Types	Illuminance (Footcandles)
	Group 1 Tasks	
A	Public spaces	3
B	Simple orientation for short visits	5
C	Working spaces where simple visual tasks are performed	10
	Group 2 Tasks	
D	Performance of visual tasks of high contrast and large size	30
E	Performance of visual tasks of high contrast and small size, or visual tasks of low contrast and large size.	50
F	Performance of visual tasks of low contrast and small size	100
	Group 3 Tasks	
G	Performance of visual tasks near threshold of vision	300-1000

Therefore, vibration and noise control are critical design elements for building systems.

Noise in a building can come from many sources: noise from the outside entering through the building envelope; noise generated by plumbing, HVAC, and electrical systems; noise generated by the building occupants; and noise produced by equipment and machinery housed in the building. The human ear is not equally sensitive to all frequencies of sound and noise can is perceived as "hissy" when the noise is concentrated in the higher frequencies, "rumbly" when concentrated in the lower frequencies, and "neutral" when concentrated in the mid-range of frequencies.

To be acceptable, the "loudness" of noise from mechanical systems must not prevent people from hearing sound they want to hear. Additionally, the "quality" of mechanical noise must not be intrusive or annoying. Therefore, the maintenance staff must be concerned about both the level of mechanical noise and its quality.

In occupied environments, sound is measured on the basis of the "A-scale." Since people do not hear low frequency or high frequency sound as well as mid-range frequency, A-scale sound meters adjust the measured levels in each of 9 frequency bands to give more authority to the midlevel ranges. The sound level at each of these 9 frequency bands is then "added" to yield a total in terms of *decibels, A-scale (dBA)*. Noise levels measured using the A-scale represent the filtering of sound that replicates the human hearing response. This is the most commonly used descriptor to quantify the relative loudness of various types of sounds with similar or differing frequency characteristics.

The World Health Organization (WHO), recommends that the level of noise in occupied spaces should not exceed those specified in Table 4-8.

Table 4-8. WHO Maximum Allowable Background Noise Levels

To Avoid	*Max. Noise Level and Duration*	
	dBA	*Hrs*
Hearing impairment	85-100	1
Annoyance	50-55	16
Disturbance of communication/lack of speech intelligibility	35	All
Sleep disturbance	30	8

Air Filters and Cleaners for Particulates

The first step in reducing exterior air particulates and pollutants is to evaluate outdoor air intakes and, if necessary, relocate them to reduce the potential for outdoor pollution. ASHRAE recommends that the following criteria be used for outdoor air intakes:

- The best location is on the windward side of the building.

- Intakes should be located in the lower third of the wall, with the bottom of the intake approximately 2 feet above grade. (For industrial applications, hospitals, and research facilities, where hazardous exhaust discharge is required, special care must be taken to ensure that re-entry of pollutants does not take place.)

- Outside air intakes must be separated from any potential pollutant source in accordance with Table 4-9.

Table 4-9. Minimum Air Intake Separation Requirements

Object	Minimum Distance (ft)
General building exhaust	5
Significantly contaminated exhaust or vents	15
Noxious or dangerous exhaust or vents*	30
Garage entry, automobile loading area, or drive-in queue	15
Truck loading area or dock, bus parking/idling area	25
Driveway, street or parking place	5
Thoroughfare with high traffic volume	25
Roof, grade, or other surface directly below intake	1
Garbage storage/pick-up area, dumpsters	15
Cooling tower intake or basin	15**
Cooling tower exhaust	25**

*Laboratory fume hood exhaust air outlets shall be in compliance with NFPA 45 and ANSI/AIHA Z9.5.
**This distance should be a great as possible to reduce Legionella contamination potential.

Filters are applied in air-handling systems to remove particulate contaminants and pollutants such as dust, pollen, mold, bacteria, smoke, and etc. from the air. ASHRAE Standard 52.2-2007 defines the methodology of rating and applying filters in HVAC systems by establishing *minimum efficiency rating values (MERV)* for filters and a test method for determining the MERV. Table 4-10 summarizes the application guidelines for each type of filter and establishes a range of recommended MERVs. This table can be used to select air cleaner performance required for most HVAC applications.

"Electrostatic filters," passive residential grade filters, are ranked at

Table 4-10. Filter Application Guidelines
(Adapted from ASHRAE Standard 52.2-2007)

Std. 52.2 Minimum Efficiency Reporting Value (MERV)	Approx. Std. 52.1 Results		Application Guidelines		
	Duct Spot Efficiency	Arrestance	Typical Controlled Contaminant	Typical Applications and Limitations	Typical Air Filter/Cleaner Type
20	N/A	N/A	**0.30 µm Particle or Smaller** Virus (unattached) Carbon dust Sea salt Combustion smoke Radon progeny	Cleanrooms Radioactive materials Pharmaceutical manufacturing Carcinogenic materials Long procedure surgery	**HEPA/ULPA Filters** ≥99.999% efficiency on 0.10-0.20 µm particles, IEST Type F ≥99.999% efficiency on 0.30 µm particles, IEST Type D ≥99.99% efficiency on 0.30 µm particles, IEST Type C ≥99.97% efficiency on 0.30 µm particles, IEST Type A
19	N/A	N/A			
18	N/A	N/A			
17	N/A	N/A			
16	N/A	N/A	**0.30 - 1.0 µm Particle** All bacteria Most tobacco smoke Droplet nuclei (sneeze) Cooking oil Most smoke Copier toner	Hospital inpatient care General surgery Superior commercial buildings	
15	>95%	N/A			
14	90-95%	>98%			
13	80-90%	>98%			**Bag Filters** Nonsupported (flexible) microfine fiberglass or synthetic media. 300 to 900 mm (12 to 36 in.) deep, 6 to 12 pockets.
12	70-75%	>95%	**1.0-3.0 µm Particle** Legionella Humidifier dust Lead dust Auto emissions	Superior residential Better commercial buildings Hospital laboratories Schools	**Box Filters** Rigid style cartridge filters. 150 to 300 mm (6 to 12 in.) deep may use lofted (air laid) or paper (wet laid) media.
11	60-65%	>95%			
10	50-55%	>95%			
9	40-45%	>90%			
8	30-35%	>90%	**3.0-10.0 µm Particle** Mold Spores Hair spray Fabric protector Dusting aids	Commercial buildings Better residential Industrial workplaces Paint booth inlet air	**Pleated Filters** Disposable, extended surface, 25 to 125 mm (I to 5 in.) thick with cotton-polyester blend media, cardboard frame. **Cartridge Filters** Graded density viscous coated cube or pocket filters, synthetic media **Throwaway** Disposable synthetic media panel filters
7	25-30%	>90%			
6	<20%	85-90%			
1-5	<20%	<80%	**>10.0 µm Particle**	**Not Acceptable for under ASHRAE 62.1**	**Throwaway** Disposable fiberglass or synthetic panel filters **Washable** Aluminum mesh, latex coated animal hair, or foam rubber panel filters **Electrostatic** Self charging (passive) woven polycarbonate panel filter

the low end of the MERV scale, despite claims by the manufacturers that this type of filter is both effective and reliable. The standard recognizes that these claims are, at best, exaggerated and the filters do not perform well or reliably in HVAC systems.

"Active" electrostatic or electronic filters are very efficient, with an MERV of 12-15, **when they are properly operated and maintained.**

However, the regular maintenance required to ensure proper filter performance is seldom provided. Internal filter failures (electrical breaks, etc.) and simply particulate "loading" on the attractor plates reduce filter performance and, since there is no obvious indication of these failures, poor performance can occur over long periods of time. *Therefore, unless the owner is aware of and committed to the required maintenance, this type of filter should not be used.*

ASHRAE Standard 62.1-2007 requires a filter MERV of not be less than 6 (and that filters be installed upstream of the cooling coils, which conflicts with the final filter requirements for some hospital and laboratory applications). But, in general, *to reduce particulates to acceptable levels, analysis by Lawrence Berkley Laboratories, along with other studies indicates that filters with an MERV of 8-9 are the minimum required. This same analysis indicates that, for comfort air-conditioning applications, filters with an MERV greater than 13 provide only minor improvement in indoor air quality unless the MERV is increased to at least 16.* Higher MERV levels may be required by regulatory bodies for special applications such as some hospital areas, research facilities, clean rooms, etc.

Since indoor air quality is dependent on outdoor air quality, the outdoor air must be "clean" and acceptable for dilution ventilation. However, in some locations, during at least part of the year (usually summer), ambient air conditions are seriously degraded by vehicles, power plants, and/or adverse atmospheric pressure conditions. For industrial buildings, or buildings located in industrial areas, the ambient outdoor air may be contaminated with particulates and/or gaseous pollutants that must be removed by air filtration and cleaning. This condition can exist even outside of industrial areas if buildings are located close to vehicle traffic, have loading docks, or even use combustion engine powered lawn equipment—all of which produce a wide range of pollutants that can be "sucked in" through outdoor air intakes.

To remove externally introduced VOCs and other pollutant gases or vapors, gaseous filtration is required, which involves chemical reaction between the filter media and the pollutant gas via *sorption*. Sorption is a process based on the electron forces within the pollutant molecule being attracted by similar forces on the surface of filter media beds. These forces create powerful chemical bonds by which the pollutant molecules are attached to the surface of the media, a process called "adsorption." To increase efficiency and capacity, the sorption media will be highly porous, producing surface areas often rated in thousands of square feet per ounce of media.

Gaseous air cleaning can also be accomplished by the process of "absorption" where the pollutant molecules (along with water vapor molecules) merge into the filter media, similar to liquid phase mixtures. The absorption process produces ionization, enabling chemical reaction to occur. To enhance this process, sorbents can be impregnated with specific reagents to enhance their ability to chemically react with specific gaseous pollutants.

There are no universal test methods of rating and selecting chemical filters. However, the following general guidelines can be applied:

Carbon is best applied against contaminants that are higher in concentration, heavier in molecular weight, and nonpolar and that have a higher molecular carbon content. This includes most VOCs and long molecular chain hydrocarbons. It also works best at lower temperature and humidity conditions.

Potassium Permanganate (KMnO$_4$) performs best against lower molecular weight compounds, polar compounds such as formaldehyde, and reactive inorganic compounds such as hydrogen sulfide and sulfur dioxide.

Zeolite is particularly effective as a cation exchange media. This makes it perform well against contaminants such as ammonia and other nitrogen bearing compounds. It also has a higher surface area than alumina compounds.

CO$_2$ and Odor Control

Dilution ventilation is a process by which some percentage of "dirty" indoor air is continuously replaced by "clean" outdoor air. When evaluating indoor air quality, the litmus test is how close to outdoor ambient air standards it comes.

Establishing criteria and methods of controlling minimum ventilation airflow in HVAC systems is becoming more and more complex. In North Carolina, there are numerous codes and/or standards that apply to establishing and controlling ventilation rates:

ASHRAE Standard 62.1-2007
ASHRAE/IESNA Standard 90.1-2007

State and Local Building Codes
ASTM Standard D6245-98, *Standard Guide for Using Indoor Carbon Dioxide Concentrations to Evaluate Indoor Air Quality and Ventilation*

The ventilation rates established in these standards and codes are based on normal building occupancy that contain the specified occupant density and activities that can normally be expected to take place. Whenever building materials, cleaning and maintenance materials, or specialized human activities introduce large quantities of specific contaminants into the building atmosphere, there may be occupant complaints and special ventilation measures should be considered to alleviate them.

Routine maintenance is required to ensure that proper levels of ventilation are maintained in the facility.

MOISTURE INTRUSION AND MOLD INFESTATION

In any building, the key to preventing mold in the building boils down to keeping water out and operating and maintaining HVAC systems to deal with any water vapor that does migrate inward. It is up to the owner to rapidly respond to and correct any water intrusion problem that may arise, including cleaning and drying any building materials that have become wet (or replacing them if they cannot be dried). *Water-damaged materials must be dried or replaced within 24-48 hours to prevent germination and subsequent mold growth!* Once mold growth is established, most building construction materials cannot be cleaned and replacement is the only option.

Preventing mold growth in buildings is a prime example of the old adage that "an ounce of prevention is worth a pound of cure." Preventing mold from becoming a problem is fairly simple and cheap, while remediating a building where mold has become established is both complex and expensive.

Molds are fungi, which are single-celled or multi-celled non-motile organisms that cannot make their own food. They rely on other organisms or their environments for nutrients. They will colonize any carbon-based material that supplies appropriate nutrients and moisture.

In general, fungi actively growing indoors can exist as yeasts or molds. Yeasts are single-celled organisms that reproduce by budding or simple mitosis. Molds are more complex, developing large colonies (mats

Figure 4-3. Indoor Mold Infestation

or balls) of multi-cellular filaments called "hypae." Some hypae penetrate the material that is being used for food, making removal very difficult. Others lie on the surface and develop reproductive structures when temperature and moisture conditions are appropriate. Fungi are ubiquitous, occurring in soil, in and on plants and animals, and in the ambient air both indoors and outdoors. Even in buildings without active fungal growth, total fungi spore concentrations of 3,000-10,000 spores per cubic meter of air are common.

In any building, any location providing high enough moisture content and a potential food source will support fungi growth.

Mold will grow on most surfaces if the relative humidity at the surface is above a critical value and the surface temperature is conducive to growth. The longer the relative humidity remains above the critical value, the more likely is visible mold growth; and the higher the humidity or temperature, the shorter is the time needed for germination. The surface relative humidity is a complex function of material moisture content, material properties, and local temperature and humidity conditions. In addition, mold growth depends on the type of surface.

Fully recognizing the complexity of the issue, the International Energy Agency Annex 14 (1990) nevertheless established a surface humidity criterion for design purposes: The monthly average surface relative humidity should remain below 80%. Others have proposed more stringent criteria, the most stringent requiring that surface relative humidity remain below 70% at *all times*. Although there still is no agreement on which criterion is most appropriate, *mold mildew can usually be avoided by limiting surface moisture conditions over 80% to short time periods and 70% for longer periods (even though some molds will begin to grow at a relative humidity as low as 60%).* These criteria should only be relaxed for nonporous surfaces that are regularly cleaned. Most molds grow at temperatures above approximately 40°F. Moisture accumulation below 40°F may not cause mold and mildew if the material is allowed to dry below the critical moisture content before the temperature rises above 40°F.

Potential substrates for mold growth include wood, paper, and textiles, which are common in buildings. Some molds will even grown on substrates that provide no nutrients, such as steel or fiberglass, as long as nutrients and water are present in the air or in dust settling on the molds.

The most important risk factor for mold growth indoors is excess liquid water in building materials due to water intrusion during construction or water intrusion from rain and/or water leaks after the building is occupied. In hot, humid climates (which includes most of the southeast United States), the second most important risk factor is high humidity, either in the air or in the substrate material due to infiltration of humid outdoor air and exacerbated by improperly designed or operated HVAC systems.

The best way to prevent mold growth is to prevent moisture intrusion problems and this process starts with the roof.

Every facility must develop and implement an aggressive roof-maintenance program as outlined in this text. This program would include periodic inspection of every roof to identify deficiencies. It is best to perform these inspections in each spring and fall. Inspections should be performed by individuals capable of determining not only apparent immediate problems, but also conditions that could become problems in the future. The inspections should concentrate on high-risk areas such as around roof hatches, drains, mechanical equipment, and high-traffic areas. As part of these inspections, also include roof flashings and drains, plumbing vents, etc.

In addition to semi-annual inspections, perform inspections after severe storms, repair or alterations to rooftop equipment, or re-roofing projects in adjacent roof areas. Perform repairs in a timely manner; once inspectors have identified deficiencies, a qualified roofing mechanic must make the repairs immediately.

Aside from the roof, foundations and walls are often the source of water intrusion problems. Groundwater can be kept out a building if the surrounding grade properly slopes away from the wall and adequate site drainage is provided. However, over time, the residue from plantings, ground cover (pine straw, mulch, etc.), etc. can build up to block the weep holes that drain masonry walls or even create a reverse flow condition so that water is trapped against the wall (ponding). The facility maintenance plan must include regular inspection and corrective action, as needed, to prevent these conditions from occurring. Each winter, when foliage is dormant, the exterior condition at the base of the entire perimeter wall should be examined.

Gutters overflow, downspouts become blocked with leaves or trash, and area site drains become plugged, all of which can cause water to drain down walls or be trapped against the foundation or walls. These situations provide ready conditions for water intrusion and must be routinely addressed under the maintenance program, with regular inspections and cleaning of gutters, downspouts, and site drains.

Caulking will fail due to age, moisture, sunlight, and movement. *Every caulk joint should be inspected annually and repaired as needed to maintain a watertight joint.* All caulking has a finite life, usually 5-15 years, and must be replaced at the end of that life. Therefore, the maintenance program should establish a caulking replacement schedule for all types of caulk joints in the building (and don't forget the hidden caulking that may be present, such as in horizontal flashings).

Double-glazed windows also fail. The problem here, aside from the fact that the window typically becomes "cloudy" and unsightly, is that the window heat transfer resistance is cut in half (or more!) when the seal around the space between the two glass panes fails and air and moisture enters this space. Now, surface condensation on the glass surface can become a problem in the winter if there are high internal moisture loads or if the building humidity level is kept at 50% or higher. This water condensing on the inside pane then drains downward to the sill and puddles, forming an ideal growth site for mold. Thus, the maintenance plan must provide for replacement of failed double glazing *when it occurs*.

For brick walls, the maintenance plan must include annual inspection to locate and repair cracks or repair other problems, as defined in Maintenance Procedure 2.2.1.1 (see Appendix B).

Finally, interior water leaks must be addressed. Interior water leaks can result from three basic sources:

1. Leaks from water piping (plumbing cold or hot water, plumbing drains, hot or chilled water lines used for heating and cooling, etc.)

These leaks can be caused by a pipe break, pipe freezing, corrosion, or latent material or construction defects.

2.   Leaks from backups and failed plumbing or HVAC systems (ruptured hot water heaters or water softener tanks, leaking cooling coil drain pans, etc.)

3.   Spills from dishwashers, washing machines, refrigerators, etc. or resulting from human error (somebody left the tub faucet running wide open and forget it!)

Preventing these leaks is almost impossible—as they say, "crap happens." *But, taking steps to first eliminate the leak and then to dry the building area affected within 24-48 hours is of paramount importance to prevent mold infestation.*

A plan to respond to water problems and, if required, mold clean-up, must be created so that water intrusion problems are identified and dealt with within the first 24-48 hours after occurrence, as follows:

- Any porous material (such as ceiling tiles, carpets, and sheet rock) that has been wet longer than 48 hours should be considered a likely source of molds. Some materials, such as drapes, can be laundered. Materials such as books and paper products may not be salvageable even if wet less than 48 hours. Carpets wet less than 48 hours need to be cleaned and thoroughly dried. Removal and reinstallation should be considered, if necessary. If the carpet is glued to concrete, it will not likely be salvageable. The underlying concrete retains moisture that is best dried out by removing the carpet and exposing the concrete to dry air for several days.

- Removal of standing water is a priority and should occur in the first 24 hours.

- Dehumidification by air conditioning, where feasible, or by dehumidifiers to remove absorbed excess water should be ongoing for 72 hours or until the environment is reduced to less than 50% relative humidity levels. A good target for moisture control is to control the dew point temperature indoors to between 35°F and 55°F.

- Sanitizing hard surfaces after water remediation or with mold growth is in evidence can be done using chlorine bleach mixed at one cup per gallon. Surfaces should be wiped with the sanitizer and left wet for 20 minutes prior to thorough drying.

In some cases, water damage may be hidden and go on for months prior to discovery (i.e., leaking water pipe in a wall, a basement crawl space damp from rain and ground water seepage, or a roof leak between the ceiling and the upper roof). In these cases significant mold growth may occur and require that the affected areas be closed down, sealed off, and decontaminated by trained staff.

A complete remediation plan for these situations should be part of the response protocol created by the interested parties.

Anytime mold is found, clean up and mitigation must begin immediately (within 24-48 hours). The two following resources are the de facto "standards" for mold mitigation and should be used to guide mold clean-up activities:

1. *Guidelines on Assessment and Remediation of Fungi in Indoor Environments*, available from the New York City Department of Health and Mental Hygiene as a free download at http://www.NYC.gov/html/doh/html/epi/moldrpt1.shtml.

2. *Mold Remediation in Schools and Commercial Buildings*, U.S. Environmental Protection Agency, EPA Publication 402-K-01-001, available as a free download at http://www.epa.gov/mold/mold_remediation.html.

Basic guidelines for remediation of water-damaged and/or mold infested materials after 48 hours, as recommended by *Mold Remediation in Schools and Commercial Buildings*, U.S. Environmental Protection Agency, are summarized in Table 4-11.

These remediation guidelines are based on the size of the affected area to make it easier for remediators to select appropriate techniques. The owner must then use professional judgment and experience to adapt the guidelines to particular situations.

In hot, humid climates, outdoor air introduced for ventilation or by infiltration, coupled with a lack of humidity control by HVAC systems, are often significant contributors to mold growth in facilities and must be addressed by the maintenance staff. But, the maintenance staff, can minimize design failures by paying careful attention to two basic factors:

Table 4-11. Remediation of Water Damaged/Mold Infested Materials

Element	Component	Requirement/Criterion
Keeping water out	Roof	Regular inspections and repairs
	Site drainage	Remove landscaping buildup around walls
		Keep drains and wall weeps clear
		Clean gutters and downspouts
	Caulking	Annual inspections and repairs
	Windows	Replace failed double glazing
	Masonry walls	Grout hairline cracks
		Tuck-point mortar joints
		Check/open weeps
		Replace broken/spalled bricks
		Check/repair dampproofing
		Inspect/repair flashing
Respond to water intrusion	Porous materials	Clean or remove within 48 hours
	Standing water	Remove within 24 hours
	Dehumidification	Maintain 50% RH or lower for 72 hours
	Remediation	Follow NYC or USEPA guidelines
Building positive pressure	Airflows	Check all airflows annually
	Exhaust air	Adjust to minimum levels
	Ventilation air	Adjust annually
Humidity control	HVAC equipment	Routine/preventative maintenance

*Clean-up Methods:

Method 1: Wet vacuum (in the case of porous materials, some mold spores/fragments will remain in the material but will not grow if the material is completely dried). Steam cleaning may be an alternative for carpets and some upholstered furniture.

Method 2: Damp-wipe surfaces with plain water or with water and detergent solution (except wood —use wood floor cleaner); scrub as needed.

Method 3: High-efficiency particulate air (HEPA) vacuum after the material has been thoroughly dried. Dispose of the contents of the HEPA vacuum in well-sealed plastic bags.

Method 4: Discard – remove water-damaged materials and seal in plastic bags while inside of containment, if present. Dispose of as normal waste. HEPA vacuum area after it is dried.

(1) maintain the building at a positive pressure at all times and (2) operate HVAC systems to provide the best humidity control that they can.

Infiltration of hot, humid outdoor air into each area of the building must be eliminated by maintaining positive pressurization in all building areas at all times. Infiltration is defined as the "uncontrolled" introduction of outdoor air into a building. One obvious source of infiltration is through open doors, but infiltration can also occur through walls, through the cracks around windows and other wall openings, through roofs, etc.

For internal pressurization to be effective in eliminating infiltration, the building pressure must equal or exceed the pressure due to wind velocity. At a typical wind speed of 10 mph during the cooling months, an internal pressure *in all areas of the building* of at least 0.009 psi (0.25" wg) would be required.

Because of the vagaries of construction, it is impossible to compute the exact amount of outdoor air that must be introduced to maintain exactly 0.25" wg positive pressure in the building. Thus, *the most common approach used by designers is to determine the exhaust requirement in each area and then introduce enough outdoor air in that area to offset the exhaust, plus an additional 10-20% as a "safety factor."* **At all times, when an exterior door is opened, air should exfiltrate (exit) from the building.**

For indoor air in the summer to be at an acceptable 50%RH or less, the HVAC systems must supply their air at a 55°F or lower dewpoint temperature.

Finally, most buildings are not occupied around the clock. Therefore, the building owner must address unoccupied periods, ranging from overnight to weekends to summer vacations, to maintain low humidity level and help control mold growth. Care must be taken to ensure that all building exhaust fans are shut down and all outdoor air dampers are closed during unoccupied periods.

The recommended approach to preventing high humidity from occurring during these periods is to establish both high limit temperature and high limit humidity space setpoints, monitored by space temperature and humidity sensors, that will cause the HVAC controls to cycle the HVAC systems on to maintain these setpoints. Ideally, this would be done for each air-handler. But, because central chiller and boiler plants may have a minimum load requirement, care must be taken by the designer to ensure that enough systems are cycled on to provide this minimum load—or there is no benefit to the unoccupied period control.

Since there are no people in the building during these unoccupied periods, ventilation air is not required and the outdoor air dampers should be closed (which eliminates the majority of the moisture load imposed on the building).

In summary, Table 4-12 represents a "checklist" that can be used to define the maintenance requirements to control mold in facilities.

DEALING WITH FAILING MASONRY WALLS

Brick, concrete, and stone masonry cavity walls allow greater water intrusion and have greater resulting mold problems than other types of exterior finish.

Table 4-12. Checklist for Preventing Mold Infestation

Components Affected	Cleanup Method (s)*	Personal Protective Equipment	Area Containment Required
Total Surface Area Affected < 10 SF			
Books and Papers	3		
Carpet and Backing	1,3		
Concrete or CMU	1,3		
Hard surface, porous flooring (VCT, vinyl, ceramics, etc.)	1,2,3	Minimum (N-95 respirator, gloves, and goggles)	None
Hard surface, non-porous flooring (plastics, metals, etc.)	1,2,3		
Upholstered furniture, drapes, or other fabrics	1,3		
Plaster, GWB	3		
Wood	1,2,3		
Total Surface Area Affected 10-100 SF			
Books and Papers	3		
Carpet and Backing	1,3,4		
Concrete or CMU	1,3		
Hard surface, porous flooring (VCT, vinyly, ceramics, etc.)	1,2,3	Limited or Full	Limited
Hard surface, non-porous flooring (plastics, metals, etc.)	1,2,3		
Upholstered furniture, drapes, or other fabrics	1,3,4		
Plaster, GWB	3,4		
Wood	1,2,3		
Total Surface Area Affected >100 SF			
or Potential for Increased Occupant or Remediator Exposure Estimated to be Significant			
Books and Papers	3		
Carpet and Backing	1,3		
Concrete or CMU	1,3,4		
Hard surface, porous flooring (VCT, vinyly, ceramics, etc.)	1,2,3,4	Full	Full
Hard surface, non-porous flooring (plastics, metals, etc.)	1,2,3		
Upholstered furniture, drapes, or other fabrics	1,3,4		
Plaster, GWB	3,4		
Wood	1,2,3,4		

With brick, the most common masonry cladding, the best design and detailing combined with the best quality materials will not compensate for poor construction practices and workmanship, which have become the norm. Proper construction practices, including preparation of materials and workmanship, are essential in attaining a water-resistant brick masonry wall. Therefore, specific construction procedures, recommended by the Brick Industry Association (BIA), must be specified by the designer and followed by the contractor to ensure quality masonry construction and avoid water problems.

As discussed in Chapter 3, however, even the best masonry wall will begin to fail—cracked or spalling units, cracks and separation of mortar joints, failed caulking, are common occurrences. Less common, but a problem in multistory brick buildings is long term creep and that results in structural failure of masonry veneer.

No matter what the cause, a failing masonry mall is difficult to address and it can almost never be repaired well enough for continued satisfactory performance. The ultimate method of dealing with a failing masonry wall is to remove the masonry material, including the existing flashing, etc., and correctly install new. However, this is an expensive undertaking and there are alternatives.

Clear water-repellent coatings are sometimes recommended to reduce water absorption and reduce the amount of water that penetrates the exterior brick masonry. However, clear water repellents can seldom stop water penetration through cracks over 0.02 in. in size. Clear water repellents cannot stop water penetration through incompletely filled mortar joints or from sources such as ineffective sills, caps or copings. Their effectiveness under conditions of wind-driven rain is questionable. *As a result, the use of clear water repellent coatings to eliminate water penetration in a wall with existing defects is often futile.*

Steel panels, using pre-finished single thickness formed sections, have successfully been applied over existing masonry walls that were performing poorly, Panels are installed on hat channels and serve as rain barriers to eliminate the direct impingement of blowing rain onto the masonry. This approach can be a low cost, though limited life, solution to leaky masonry wall problems.

A short-term (5- to 10-year) solution to deteriorating or poorly designed or construction masonry walls is to apply an elastomeric acrylic or silicone masonry coating. These opaque coatings, with an applied thickness of 8-12 mils (about 50-100 sq. feet/gallon), are applied by rolling or spraying to form a flexible, textured or smooth finish. Elastomeric wall coatings are formulated to stretch and bridge cracks (assuming proper crack preparation and sealing) that may form due to movement. When the wall contracts as temperatures drop and the cracks expand, the elastomeric wall coating stretches and resists cracking itself. When the temperature rises and the cracks become smaller, the coating returns to its original shape without wrinkling or leaving ridges. This crack bridging capability results in a long lasting, uniform finish.

To achieve the balance of properties required to make a top quality elastomeric wall coating, manufacturers use acrylic binders that are designed specifically for this use. Using these acrylic binders provides the coating with a high degree of flexibility and elasticity, while exhibiting appropriate tensile strength, and superior alkali resistance and dirt resistance, which are important considerations. Elastomeric wall coatings cure to a smooth, non tacky, high dirt resistant finish which will resist cracking and wrinkling.

Good surface preparation is critical for a good coating. With masonry, certain aspects need special attention, such as alkalinity, porosity, and the tendency to form efflorescence. The masonry substrate must be clean and sound so that the coating can adhere properly to its surface. This requires water blasting or sand blasting to remove any dust, dirt, flaking paint, or other loose material. If the surface was previously painted, mildew or mold may be present. These contaminants should be removed with a common bleach solution (three parts water to one part household bleach). In all cases, after performing any surface preparation procedures, the surface should be rinsed well with clean water.

Once the surface is clean, cracks greater than 1/16 inch should be sealed with a top quality water based acrylic or silicon-acrylic caulk or sealant. Furthermore, if the surface remains somewhat chalky or is porous, a water- or solvent-based masonry sealer must be applied before applying coating. If the surface tends to form efflorescence, or is new and highly alkaline, a masonry sealer or primer is required to protect the textured coating elastomeric topcoat.

Textured coating elastomeric wall coatings will perform well only when applied in thick films. The typical application procedure is to use either an airless spray followed by back rolling, or a long nap roller. A 1-1/2" to 4" brush works well for smaller areas.

Properly formulated coating elastomeric wall coatings have a satin finish. They can generally be tinted and while white is most popular, soft pastels and subtle earth tones are often chosen as decorative options.

Ideally, once a textured or smooth coating elastomeric coating is applied, it should be allowed to weather for at least one year before a non-elastomeric coating is applied over it. Even then, surfaces with coating elastomeric coatings should be repainted only with a water-based topcoat, since oil-based coatings are much less flexible, and can crack severely if applied over the elastomeric coating.

Finally, another, but often maligned, product that can be successfully applied over masonry walls to seal them is *externally finished insulation systems* (EFIS). In fact, EFIS was originally developed in Europe specifically to be used on masonry construction. (It was only after being widely, and poorly, applied to wood frame construction in the United States that the water problems with EFIS became evident.) Drainage plane EFIS, properly applied over masonry, can be a permanent (30-year) solution to wall moisture intrusion problems. However, the following potential problem areas must be addressed:

- Ensure that existing caulking and flashing is properly installed or, if not, is properly repaired.

- Detail all EFIS intersections—roof overhangs, windows and doors, grade level, etc.

- Ensure that the installer is licensed and trained by the EFIS manufacturer and that he uses only that manufacturer's products and installs them properly.

Chapter 5

Designing for Building Maintainability

Why do architects and engineers spend four or five years in college addressing design issues and practically no time on building maintenance? In practice, there is still only a limited degree of association and relation between maintenance professionals and designers. Designers continue to create many buildings which fail to achieve satisfactory long-term performance and maintainers still have little evident influence or impact on design. Design and maintenance issues require a practical approach, a problem solving orientation and a "learning from experience" to produce a comprehensive approach to improving building performance and reducing life cycle cost.

Unfortunately, many factors relating to design and maintenance create contrasting and contradictory objectives and in accommodating certain design needs, maintenance requirements may be adversely affected. Conversely, the attempt to produce maintenance free buildings may impose completely impractical and certainly uneconomic design provisions. Consequently, there is a need to compromise and balance the opposing criteria to improve overall building performance.

While we know that the quality and performance of buildings is dependent on the effectiveness of its design and sustained maintenance, traditionally, these tasks are treated as separate entities. This intrinsic separation causes several problems such as:

- Lack of feedback and support: Design and maintenance professionals often have minimal knowledge or respect for each others problems. The priorities considered in design frequently rank maintenance as the lowest need and the lessons learned through remedial measures rarely influence design procedures.

- Lack of knowledge and understanding among design professionals: Due to the inherent separation and under the pressure of time/budget, design consultants often do not have sufficient understanding and knowledge on maintenance issues during office building design stage. As a result, there are limited considerations for service, upgrade and maintenance during the life span of the building.

• Lack of data on operational requirements: There is a lack of historical data concerning the operational requirements and maintenance performance of existing buildings. As a result, designers have to work with insufficient and within the limited times, even inappropriate information.

• Difficulties in forecasting future condition and changes: Increasing global warming may have a significantly important effect on the climate conditions for buildings to withstand. Without necessary support, designers are unable to predict such changes for remedial measures for building servicing and maintenance.

• The "First cost" mentality: Influenced by owners and developers, designers often focus on the initial cost of building, i.e. built cost, without sufficient consideration for subsequent costs such as that incurred in building maintenance. Capital cost is only 20% of the total cost of a building over its lifespan. How buildings will perform or what they will cost to run 10 to 30 years in the future is not a key issue considered by most developers.

Additionally, designers are subjected to a great deal of lobbying to use one product rather than another, and are bombarded with an overabundance of publicity material. This can have two severe consequences:

• It may influence the designer to use a new component (material, system, etc.), which has not been sufficiently proved in practice, with a consequential failure to meet requirements.

• Conversely, there may be the opposite reaction, resulting in the designer using a component or material with which he is familiar, whether or not it is the right choice in the circumstances.

A major short-coming in the design process appears to be a failure to make use of the authoritative guidance that is available. This might be a simple failure to use codes and standards properly, or may be attributable to the shear volume of advice. Another source that many designers (and most contractors) overlook is manufacturer installation instructions, resulting in poor performance of the product that is installed incorrectly (think Tyvek barrier without taped joints).

And, most importantly, without using life cycle costing for design decision making, buildings will lack thorough documentation to reveal the real costs during the life of the building...designers focus on the construction cost of a building, but typically do not really take into account the subsequent owning and operating costs, especially maintenance costs.

Some effort has been attempted to facilitate improved communication and management to create a better balance of design and maintenance. One approach to this is for a single group made up of the designers, the contractor(s), and the owner's operations and maintenance staff to coordinate the design effort...what we now call the "integrated design" approach. Another alternative is to forcefully involve greater participation by assigning legal responsibilities to designers and developers.

However, attempts to facilitate effective communication during the design process are burdened by conflicting objectives, professional prejudices, and inappropriate management structures. If design and maintenance issues are to be balanced, it is essential for one group to coordinate design, construction and maintenance activities and much work remains to ensure that the integrated design approach becomes widely used.

For this reason, some countries (Singapore, for example) require developers to retain at least a 30% share in the building for ten years after completion, automatically involving them in operations and maintenance issues, while others (such as Taiwan) require developers to handover a maintenance budget to the building management control board during the building delivery process.

While these compulsory requirements can actively assist in increasing maintenance assessment, they are resisted by developers and few governments have considered them. *Thus, improved design and construction techniques, ideally determined and incorporated via the integrated design process, remain the best way to achieve good quality buildings.*

Over the whole life of a building, the design and construction period represents only 1-10% of the life cycle; the rest, including operation and maintenance, makes up 90-99%. Under ideal circumstances, a building's design would represent a perfect model of the proposed facility which, as well as being of relevance to the detailed design and construction, will be of value to effective maintenance of the building.

Building studies in the United States, the U.K., Australia, Singapore, Canada, Taiwan, etc. have revealed that the majority of building and/or component failures (or, at least, high maintenance requirements and/or shortened performance life) result from making wrong choices about materials or components for a particular situation in the building. *Thus, maintenance must be fully considered at the design stage.* Each of the following design elements is critical:

- The adequacy of the design and the suitability of the materials specified

- The standard of workmanship in the initial construction and subsequent maintenance operations...an "oversight" responsibility of the designer.

- The extent to which the designer has allowed for present and anticipated future needs

GENERAL CONSTRUCTION

Based on research at Harvard University, building quality and resulting maintenance requirements imposed on the building owner result from decisions made by the designers in the earliest stages of the design process. Table 5-1 summarizes Harvard's analysis that each major component of the building design must be considered relative to service life and maintenance.

Most commonly, general construction maintenance problems relate to moisture intrusion and/or air infiltration resulting from improper design and/or construction.

Exterior Walls

Failing masonry walls, as discussed in Chapter 4, are a prime culprit for moisture intrusion, resulting in the need for ongoing and, ultimately, expensive, maintenance. The basic way to limit potential moisture intrusion problems is to avoid the use of masonry exterior walls.

Table 5-1. Quality of Building Components vs. Design Service Life

Item	Level 1	Level 2	Level 3	Level 4
Service Life	100 years (Museums, Performing Arts, Monumental Bldgs.)	50 years (University Bldgs., Gov't Bldgs., etc.)	30 years (K-12 Schools, Office Bldgs., Industrial, etc.)	15 years (Retail, etc.)
Space Efficiency	60%	70%	75%	80%
Super Structure				
Floor Heights				
First	15 to 25 feet	15 to 20 feet	15 feet	12 feet
Other Floors	14 to 15 feet	13 feet	12 feet	11 feet
Spans	35 to 40 feet	30 to 35 feet	25 to 30 feet	20 to 25 feet
Live Load	100 psf	100 psf	80 psf	50 psf
Min Stair Width	6'-0"	5'-0"	4'-0""	3'-8"
Wall Construction				
Material Quality	Granite/Marble	Limestone/Brick	Precast Concrete /Brick	Curtain Wall/Wood/Vinyl/CMU
Fenestration Type	Bronze	Anodized Alum	Coated Alum/Steel	Mill Finish Alum
Glazing	Insulated Low E	Insulated Low E	Insulated Tinted	Insulated Clear
Insulation R-Value	R-30	R-30	R-20	R-16
Roofing				
Material Quality	Slate/Tile/Copper	5-ply BUR/Metal	Single Ply, Modified, 5-ply BUR	Single Ply
Insulation R-Value	R-30	R-30	R-19	R-19
Interior Construction				
Lobby Floors	Stone	Terrazzo	Quarry tile	VCT
Corridor Floors	Terrazzo	Terrazzo	Carpeting	VCT
Partitions	Plastered Masonry	Sound Resistant Metal Stud and Drywall	Metal Stud and Drywall	Metal or Wood Stud And Drywall
Ceilings	Plaster/Acoustic	Splined/Acoustic	Acoustical Panel Tegular Grid	Acoustical Panel Flush Grid
Conveying Systems				
Elevators	Traction	Traction	Hydraulic	Hydraulic

Also, experience shows that wood as a exterior wall material has significant maintenance issues (wet rot, dry rot, etc.) and should be avoided, especially in humid climates. *Eliminating wood and masonry as wall cladding will significantly reduce exterior wall maintenance problems.*

What materials should be used as the cladding of exterior walls? First, consider the use of precast concrete panels, which have approximately the same design service life as masonry materials, but with far fewer maintenance problems. Precast concrete panels are denser than brick or CMU and, thus, are better as water barriers. Installation is more "engineered" and less dependent on the workmanship to produce a wall without leaks, cracks, settlement, etc. that require maintenance. And, from an aesthetic perspective, precast panels can be formed in many different shapes and exterior finishes, yield significant design freedom in their application.

Another good choice is prefinished, insulated metal panels. Depending on metal thickness and type of finish, metal panels can have an anticipated service life as great as precast concrete, though typically will have a life expectancy about 30% shorter. Again, installation is more "engineered" and less dependent on the workmanship to produce a wall without leaks that require routine maintenance.

Finally, shorter life materials that should be considered include vinyl, cement board, and metal cladding. These materials perform well and require little maintenance over their design service life. And, at the end of that life, they are relatively simple to replace.

The fundamental concept in rainwater control for wall design is to shed water by layering materials in such a way that water is directed downward and out of the building. Gravity is the driving force behind water drainage. The "down" direction harnesses the force of gravity and the "out" direction gets the water away from the building. In general, the sooner the water is directed out, the better.

Wall design must be based on the assumption that water can penetrate the exterior finish or cladding and enter the wall.

Every wall, therefore, must have a rain or water barrier applied to the outside surface of the backing wall to ensure that wind-driven rainwater does not penetrate the inner wythe of the wall. The exterior cladding and be installed in such as way as to allow water to drain to the exterior of the building. Then, a cavity or "drainage plane," with proper through-wall flashing and drain outlets, is always necessary. Water that penetrates through cladding, then, will fall by gravity down the drainage cavity and drain to the outside, not to the inside of the building.

Fenestration

Fenestration represents a "hole in the wall" through which water, water vapor, and air can enter a building. Windows, even the very best, do not have a thermal performance that even approaches that of a well-insulated wall; have you ever seen an R-30 window? So, unless very carefully designed, almost all fenestration represents increased energy inefficiency for the building envelope. *To minimize these negative impacts and the maintenance associated with dealing with them, the quantity of fenestration should be held to the minimum required.*

Next, avoid operable windows. Operable windows typically result in increased infiltration and resulting moisture intrusion, and tend to leak over time, requiring routine maintenance. Fixed glazing systems avoid many of these problems.

But, even with fixed glazing, careful detailing of the window installation is required to prevent water intrusion. Both the tops and bottoms of windows must be flashed. Flashing over the window top trim is required. In all cases, windows should be recessed in the wall to provide a drip edge along the top and/or provided with a specifically designed drip edge. Pan flashing is a highly recommended method for creating an under-window "gutter" to redirect any leaking water to the exterior.

The problems with window-wall systems are even more complex. Unlike traditional curtain wall systems that span floor to floor with attachments outboard of the slab edge, the traditional window-wall system is designed to bear on the slab edge and extend up to the underside of the slab edge above. In some systems, a slab band cover is provided to give an appearance similar to curtain wall. Although there are many variations, the typical window-wall system is defined by the following features:

- Window extrusions used to span from floor to floor in a pre-manufactured unitized system typically up to 6 feet in width.

- A coupler connecting the vertical mullions of each window-wall unit, either as a male-female coupler integral to the vertical mullions or as an independent coupler.

- Window-wall systems bear on the slab edge at the floor level and are anchored to a retaining track fastened into the slab.

- The head of the window-wall system is slotted into a deflection header fastened to the underside of the floor slab, spandrel beam, or structural wall and anchored to the structure with deflection straps.

- Horizontal mullions dividing the window-wall unit into panels of vision glass units, glazed from either the exterior or interior, and glass or metal opaque panels mounted from the exterior.

- Operable units, typically awning or casement windows, swing doors, and sliding doors, mounted within the frame of the window-wall system.

- A slab band cover spanning the depth of the slab that can be installed either independent of the window-wall system or as an extension of the window-wall frame.

Early window-wall systems typically used a face sealed approach to their design with respect to water tightness, relying on the continuity of the primary exterior seal to prevent water infiltration. Many of these systems have experienced chronic water infiltration problems as the result of deficiencies in the primary exterior seal. These were often aggravated by air leakage that emphasized the deficiencies in the primary exterior seal under differential pressure. Once the water is past the exterior seal, it often accumulated within the frame, as there was no provision to contain this water or to drain the interstitial cavities. Eventually, the accumulating water overflowed to the interior.

These frequent water ingress problems have led to the adoption in recent years of rain-screen design concepts. The adaptation of rain-screen concepts to window-wall systems was twofold. In the first instance, a second line of protection against water penetration was provided inboard of the original primary exterior seal. Different assemblies within a typical window-wall system require different approaches to providing an effective second line of protection, including

- Installation of a secondary seal for the vision units, typically provided by a heel bead of sealant applied between the inside light of glass and the frame of the window-wall system for interior glazed units;

• Installation of a metal back-pan sealed to the frame of the window-wall system behind all opaque panels to act as a secondary seal;

• Installation of a waterproof membrane at the level of the floor slab extending from the vertical leg of the mounting track down the face of the slab and onto the deflection header.

The primary intent of these measures is to provide redundancy within the system. The seals forming the secondary line of protection are typically better defense from exposure and temperature extremes than the primary exterior seals and can be expected to have a longer service life. In the second instance, means of draining the interstitial spaces created between the primary and secondary seals are provided so that any water making its way past the primary exterior seal would not accumulate within the window-wall system but would be redirected to the exterior. To reduce the risk of water accumulation within the system, designers must ensure that each element of the window-wall is provided with means of draining, including

• Drainage of the glazing cavity and back-pans, either directly to the exterior by way of drain holes through the horizontal mullions or down to the deflection header by means of drain holes in the vertical mullions;

• Drainage of the cavity between the primary and secondary gaskets on operable units;

• Drainage of the horizontal mullions directly to the exterior;

• Drainage of the window-wall system at the level of each deflection header.

An additional advantage of this twofold approach is that it creates a series of compartmentalized cavities that provide for some measure of pressure moderation across the primary exterior seal. Reducing the pressure across the exterior seal reduces the amount of water that can be driven through the minor imperfections likely to be present in the exterior seal. The compartmentalization of the window-wall system also makes it easier to isolate any eventual leaks. With the earlier window-wall systems, water might find its way down a few stories from its point of entry before any sign of water ingress was visible on the interior. With a compartmentalized system, it is unlikely that water can travel far from its point of entry.

Roofing

All roofs will leak; it is just a matter of when. But, pitched or high-sloped roofs shed water better than flat or "low-slope" roofs (roofs with less than 3:12 slope) simply because gravity *always works* and roof design and construction may not. High slope roofs (with pitch greater than 3:12) have the advantage of shedding rainwater and reducing the potential of roof leaks into buildings since the roof doesn't actually have to be a water barrier. *A roof slope of at least 3:12 is highly recommended*, the result will be a roof that lasts 2-3 times as long, leaks far less, and significantly reduces the cost of maintenance.

As part of a high slope roof system, the use of an architectural standing seam metal roof (ASSM) as the primary rain barrier is becoming standard. Asphalt shingle roofs wear poorly in almost all climates, wood shingle roofs are a fire hazard and maintenance nightmare, and tile, cement, and slate roofs are too expensive and too heavy to ordinarily considered. ASSM is typically made of galvanized steel with an aluminum coating, covered with a baked-on enamel or powder-coat finish, has a long design service life, and requires very little maintenance.

If, however, after a thorough evaluation, the decision is made to utilize a low slope roof, the next step is to decide the type of low slope roof to install. Roof slope should never be less than 1:50 (approximately 1/4" per foot) and the slope must be uniform to prevent "ponding." There are basically two types (with many variations) to consider:

Type 1

Single-ply materials that can be attached chemically, mechanically, or held in place with ballast, usually gravel or aggregate. Examples of single-ply products include:

— EPDM: Ethylene propylene diene monomer (synthetic rubber) is a flexible elastomeric material.

- — PVC: Polyvinyl chloride a synthetic thermoplastic polymer prepared from vinyl chloride.
- — TPO: a blend of polymers that may or may not contain desirable additives such as flame retardants or UV absorbers.

Type 2

Built-up roof (BUR), consisting of multiply plies of two components, bitumen and felts. Bitumen provides the waterproofing and adhesive properties of the system. The felts strengthen and stabilize the membrane. The roofing membrane is protected from the elements by a surfacing layer, either a cap sheet or gravel embedded in bitumen or a coating material.

While there are many pros and cons for each of these two types, *installed correctly, a 5-ply built-up roof is far better than any single-ply membrane in terms of life cycle performance and ease of maintenance.* The BUR has many advantages:

- With a gravel wear surface, the roof has excellent exterior fire resistance.

- It is impact and puncture resistant—better for roofs with high foot traffic.

- Temperature differential stress due to water ponding is significantly lower.

- Repairs are fairly easy (and can often be done by the facility maintenance staff)—no special manufacturer licensing requirements for contractors.

- Multiple plies add redundancy for water barrier protection—a 5-ply roof is better than a 3-ply roof.

- Service life—which can be extended by ongoing maintenance and repair.

"Cons" associated with BUR include odors and potential fire hazards during installation, it is susceptible to damage from animal fats and solvents around kitchen exhaust fans, coal tar is susceptible to cold weather cracking (bitumen is not), and more complex details are required for roof penetrations and at parapets and roof curbs.

With any type of low slope roof, the following elements must be considered:

1. The use of a non-hygroscopic insulation is highly recommended, since this type of insulation, even if wet, will retain most of its thermal properties. The two most common non-hygroscopic insulation materials are cellular glass (expensive, but well worth it if significant roof traffic is anticipated) or polyisocyanurate. If roofing insulation gets wet, and if it is the type of insulation whose thermal performance degrades with moisture content (fiberglass, mineral wool, etc.), conditions are such that condensation in or on the bottom of the roof can occur. This will allow both mold growth and condensation that can drip onto ceiling structure, tiles, ducts, and pipes below to contribute to mold growth there.

2. Consider locating rooftop equipment inside penthouses. If equipment is located outside on the roof, a platform or continuous roof curb is required. Platforms should be structural steel frames above the roof surface. Pipe columns through the deck to the structural frame are the simplest to flash. *Clearance below the equipment should be sufficient to allow for roofing service and repairs.* Structural roof penetrations must be sealed, flashed, and counter-flashed to prevent water intrusion. Cubs must be full perimeter type, designed for the type of roof. Curbs must extend a minimum of 8" above the roof surface or above the height of any emergency overflow pipe or scupper, whichever is higher. Curbs must be fully flashed and have detachable secondary counter-flashing around all four sides.

    Provide walk pads to and around any equipment requiring routine service or inspection to protect the roofing membrane.

    Piping and ductwork routed across the roof must be supported from engineered roof rails with the roof membrane extending up under a metal counter-flashing. Rails must extend a minimum of 8" above the roof surface or above the height of any emergency overflow pipe or scupper, whichever is higher.

3. A vapor retarder is a low slope roof component designed to restrict

migration of water vapor molecules from areas of high pressure to areas of low pressure. In a cold climate, this vapor migration is in an upward direction. In a hot, humid climate, the vapor migration is reversed and is in a downward direction, i.e., into the building.

For a material to qualify as a roof vapor retarder, its vapor permeance rating should not exceed 0.1 perms. Materials qualify as vapor retarders include polyethylene film, PVC (polyvinyl chloride) sheets, aluminum foil, coated Kraft paper or laminated Kraft paper with bitumen filler, saturated felts (2-plies) with two or three asphalt moppings, and coated base sheet with one or two asphalt moppings.

There are advantages to having a vapor retarder:

a. A vapor retarder can ensure the continual thermal resistance of insulation sandwiched between the vapor barrier and built-up membrane.

b. A vapor retarder is a good safeguard against vapor migration if a building's use changes from "dry" to a "wet" use. (Typically "dry" is defined as an interior winter relative humidity under 20 percent, which almost never happens in hot, humid climates, while "wet" is defined as an interior winter relative humidity over 45 percent, which is common in hot, humid climates.)

c. A vapor retarder is advisable over wet decks (poured gypsum and lightweight insulating concrete) to prevent vapor flow upward into the insulation. For this purpose, the vapor retarder should be a venting base sheet, with mineral granule underside surfacing, mechanically fastened to the decks.

And, there are disadvantages:

a. The vapor retarder, together with a roofing membrane, inevitably seals trapped moisture within the roof "sandwich," which can eventually destroy the insulation, split or wrinkle the built-up membrane, or in gaseous form, blister it.

b. In the event of a roof leak through the membrane, the vapor retarder traps the water below the insulation and releases it through punctures that may be some lateral distance from the leak, making the leak difficult to locate. A large area of insulation may be saturated before the punctured membrane can be repaired.

c. A vapor retarder is a disadvantage in summer, when vapor migration is generally downward through the roof. Hot, humid air can infiltrate the roofing sandwich through the vents, or by diffusion through the membrane. It may also condense on the vapor retarder itself.

d. A vapor retarder may be the weakest horizontal shear plane in the roofing sandwich. Failure at the vapor retarder-insulation interface can split the membrane. The vapor retarder introduces an additional component where shear resistance may be critical to the membrane integrity.

When should a vapor retarder be used? The old school of thought was: "If in doubt, include a vapor retarder." However, modern policy is: "If in doubt, omit the vapor retarder." ASTM Manual 18, *Moisture Control in Buildings*, recommends, based on research by the Army's Civil Engineering Research Laboratories, consideration of a vapor retarder on any roof where the indoor winter relative humidity at 68°F exceeds approximately 60% RH (ranging from 55% RH in the western counties to 70% RH in the extreme southeastern counties). At 72°F indoor temperature, a more typical winter condition, this criteria is equivalent to a 70% RH. Facilities with such high indoor winter humidity levels may include some research facilities, laundries, cafeteria/kitchens, natatoriums, athletic facilities, "process" plants, dorm shower areas, etc.

When a vapor retarder is not needed, it should not be used since it is expensive and, more importantly, will allow "cancers" of wet insulation to grow within a compact roof system having membrane or flashing flaws. Flawed roofs without vapor retarders tend to leak water into the building sooner, which often reduces the lateral extent of wet insulation. And, typically, little guidance is provided to designers as to how to seal vapor retarders at flashing and penetrations. If a vapor retarder is required, the designer must research and carefully detail these conditions.

Research shows that roofing insulation, once wet, takes a year or more to dry if there is no vapor retarder. This insulation *never* dries if there is a vapor retarder to trap the moisture.

Other vapor retarder design considerations should be incorporated:

a. A built-up roof application does not require a vapor retarder, as the roof membrane serves a dual function as water barrier and vapor barrier.

b. If a single-ply roof application requires a vapor retarder, it should be located between the insulation and membrane. If a single-ply roof application is installed over an existing built-up roof, a vapor retarder is not required, as the existing built-up roof will serve as the vapor retarder.

4. Roofs and flashings normally require "blocking" for vertical and/or horizontal alignment or for close off. Typically, the blocking specified is treated wood. Alkaline copper quat (ACQ-D) and copper azole (CA-B) are the two primary preservatives that have replaced chromated copper arsenate (CCA) in pressure-treated wood products. Recent research shows that these new preservatives are 2-5 times more corrosive to galvanized steel fasteners, because of their high concentrations of copper, due to galvanic reaction. Highly galvanized (G-185) hot-dipped or powder-coated fasteners and connectors must be used to prevent premature connector failures that result in roof leaks.

5. Flashing design should comply with *Architectural Sheet Metal Manual*, Fifth Edition, Sheet Metal and Air Conditioning Contractors National Association (SMACNA). Galvanized steel flashing, *unless hot-dipped galvanized after fabrication, should not be used in hot, humid climates because of problems with corrosion (especially with treated wood blocking or near salt water) and painting*. It is always better for maintenance to use copper, stainless steel, or lead for flashing; do not use flexible membranes.

Interior Construction and Finishes

There are two considerations relative to interior construction and finishes, the first is design life and maintainability, while the second relates to some finishes, such a vinyl wallcovering, oil-based paint, etc., that can act as vapor retarders.

Gypsum wallboard (GWB) is a common interior wall material. But, GWB is much easier to damage than concrete masonry units (CMU) and should not used to enclose high use areas such as corridors in schools, hospitals, etc. Interior doors should always be the solid core type; hollow-core doors are far too easy to damage. Designers should specify quality door hardware to reduce routine maintenance.

Design life for an interior finish is generally considered to be its "wear" life. Floor and wall coverings have relatively short service life (3-7 years) and, thus, represent a significant maintenance burden. To reduce this burden, designers must consider wear when selecting materials, especially in areas of high use.

For wear, the best floor finishes are terrazzo, stone, tile, and other hard finish masonry materials. In the middle there is vinyl tile and epoxy-coated concrete. The poorest wear finishes include carpet and wood. Application of each of these types of materials must be based on a careful evaluation of the wear load to be imposed and detailed life-cycle cost analysis. Some areas are a no-brainer; if carpet is used in elevators, corridors, lobbies, etc., then carpet squares imposed markedly lower maintenance requirements. In wet areas, tile, epoxy-coated concrete, etc. may be required.

Inappropriate wall finishes can create maintenance nightmares. For example, in hot humid climates, vinyl wallcovering installed on a exterior wall will act as a vapor barrier, trapping moisture migrating through the wall from outdoors and creating a breeding ground for mold growth. However, in cold climates, failure to use oil-based paint or vinyl wallcovering as the interior finish in kitchens, toilets, or other humid areas may result in moisture entering exterior walls, freezing, and damaging the wall construction.

PLUMBING

Two aspects of building design have major impacts on the maintainability of plumbing systems: (1) the types and quality of plumbing fixtures, especially flushing fixtures, and (2) access for maintenance and, ultimately, replacement of fixtures and other plumbing components.

The quality of flush valves (or, for residential occupancies, flush tank assemblies) has a major impact on plumbing maintenance requirements. Cheap flush valves have high failure rates and short lives that impose significant maintenance burdens (in addition to wasting significant water).

The most common problem in plumbing systems is stoppages in waste flows, either at the fixture or further in the piping systems. Cleanouts are critical and every design should include cleanouts at each change in flow direction and at the ends of all branch waste mains. As on plumber says, "there are never too many cleanouts" and the code requirements tend to be woefully inadequate.

Access for maintenance is a prime design consideration. Too often water heaters are installed in totally inaccessible locations...above ceilings, in storage closets, etc. In dormitories, hotels, and other domicile occupancies, rarely are provisions made to provide access for replacement of tub drains or shower valves, an often critical need.

The location and service access for backflow preventers and pressure reducing valves must be a prime plumbing design criteria. These devices need routine testing and adjustment and often fail, requiring replacement.

And, finally, flood control measures are required wherever there is potential for plumbing water release...at backflow preventers, water storage tanks, water heaters, etc.

HEATING, VENTILATING, AND AIR-CONDITIONING

Two aspects of building design have major impacts on the maintainability of HVAC systems: (1) where HVAC equipment is located and the access provided for maintenance and, ultimately, replacement and (2) the indoor and outdoor space provided for HVAC equipment maintenance.

HVAC Equipment Location and Access

HVAC system components are typically located on the roof outdoors, on the ground outdoors, in mechanical equipment rooms indoors, and within ceiling cavities indoors. There are specific requirements relative to maintainability of HVAC equipment at each of these locations in the following sections.

Outdoor equipment is easier to install, easier to service, and much easier to replace in the future if located on the ground rather than on the roof. If a rooftop location is selected, building codes may require safety railings be installed to protect service personnel if the equipment is installed within six feet of the roof edge. If the roof slopes, level service platforms may also be required. And, ultimately all rooftop equipment must be replaced, so staging areas for cranes, etc. must be provided.

The following guidelines should be followed to ensure proper outdoor equipment operation and accessibility for maintenance:

1.   Air-cooled condensing equipment rejects heat to the outdoor air and, therefore, these units cannot be enclosed or placed in a position where the rejected heat recirculates into equipment intakes. There are three concerns relative to this equipment:

    a.   Provide space for the units, adequate separation between multiple units, and space around the units for maintenance and air movement in accordance with the manufacturer's requirements.

    b.   Locate the equipment so as not to create a "heat trap" that impairs performance of the equipment. Air-cooled equipment should be located at least half the wall height away from adjacent solid walls. Thus, if the wall height is 18 feet, the equipment must be located at least 9 feet away from the wall. The equipment cannot be located in an area defined by more than two walls of the building (an inside corner). Putting equipment in building recesses (3 building walls) or in wells (4 building walls) is guaranteed to impair the performance of the equipment, reduce its design service life, and increase maintenance requirements.

    c.   Equipment screening must have at least 50% open area and the screen height cannot exceed 1-2 feet above the height of the equipment.

2.   Cooling towers reject the heat removed from the building by the chiller(s) to the outdoor air. This heat is rejected by spraying condenser water over a large heat transfer surface area, allowing a percentage of the water to evaporate and cool the remaining water. Cooling towers may be located on the roof or on the ground. If located on the ground, they must be installed so that the operating water level in the tower basin is higher than the condenser water pump and all associated condenser water piping. This may make the overall installed height significantly higher than envisioned.

       The outdoor area requirement for cooling towers is approximately 0.5 sf per gsf of building area, plus the additional clearance space for airflow and service shown by Figure 5-1.

       Since the discharge of any cooling tower or evaporative condenser will contain a small percentage of respirable water droplets, called "drift," care must be taken to ensure that HVAC air intakes are located well away from these discharges.

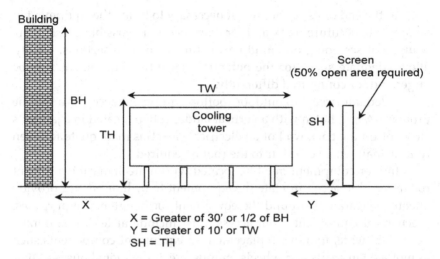

X = Greater of 30' or 1/2 of BH
Y = Greater of 10' or TW
SH = TH

Figure 5-1. Cooling Tower/Evaporative Condenser Installation Guidelines

3. Most fuel storage tanks are installed above grade and provided with a spill containment enclosure. While the tanks should be located relatively close to the boilers to reduce fuel piping and pumping costs, there is no real restriction on their location. However, screening and locked doors and gates will typically be required. Drains from the open containment area must be provided with locked valves, to be opened to drain rainwater only by authorized personnel.

4. Packaged hot water boilers designed for outdoor installation are available in capacities up to about 2,000,000 Btu/h. Typically these boilers are about 3-6 feet square and 4-6 feet high and a single boiler requires 75-100 sf of outdoor area to allow for service. Two boilers require 120-140 sf of area, with each additional boiler adding 50-60 sf to this requirement. The location, arrangement, and screening requirements for outdoor boilers are generally the same as for air-cooled equipment. However, since the hot surfaces of these boilers represent a potential public safety hazard, access to this equipment must be controlled via secure screening with lockable access openings.

Outdoor ground level equipment access for routine maintenance is rarely a problem. However, when time comes to replace the equipment at the end of its service life, it necessary to locate the equipment in an area where future truck and/or crane access is possible. Design any equipment screening or sound barriers in removable sections. Finally, allow adequate space for the potential that future equipment may be larger and/or configured differently.

Adequate access to outdoor rooftop equipment access is a routine problem. A roof hatch with a vertical ladder (often located in a janitor's closet or on the back wall of an electrical closet) is inadequate for even routine maintenance; a stair to the roof is required.

Indoor equipment must be located within mechanical equipment rooms sized to accommodate the equipment to be housed and provide adequate clearances around the equipment for maintenance purposes. Mechanical equipment room access must include at least one double door, 5'-6' wide, to allow replacement of equipment components such as motors, fan shafts and wheels, pumps, etc. If the room houses a chiller, removable wall panels or even roll-up doors should be provided so chiller tubes can be cleaned and replaced and the chiller itself can be replaced in the future.

Ceiling cavity space should not be used to house any HVAC equipment except for terminal units or small fancoil units. Ceiling cavity access to HVAC equipment can be provided by using a removable panel (lay-in) ceiling. If the ceiling is not accessible, access panels or doors are required to provide for routine service of terminal units, fire and smoke dampers, duct-mounted coils and humidifiers, controls, etc. If the equipment contains a fan and/or a refrigeration compressor, provisions for removal and replacement in the future must be provided.

Mechanical Equipment Rooms for Air-Handling Equipment

While smaller buildings may use outdoor rooftop air-handling units, larger buildings will require "fan rooms" to house this equipment. The fan room dimensions are based on the size of the air-handling unit and its required maintenance space. The length (L) and width (W) of a typical air-handling unit required to serve a given building area can be determined from manufacturer data and, then, the minimum room dimensions established as follows:

Room Length = 1.25 x L
Room Width = 2.5 x W
Room Height = 12'-14' (minimum to underside of structure)

On construction documents, designers should show coil pull, filter pull, and motor access space requirements to ensure that piping, raceway, or other equipment does not intrude into these spaces.

Designers must ensure that flood control is provided in all fan rooms; a single, uncontrolled water leak can do hundreds of thousands of dollars worth of damage to a building. All floor openings should be protected with 2"-4" high curbs or pipe sleeves to "dam" any loose water on the floor. Adequately-sized floor drains must be provided, with floors sloped to these drains.

Mechanical Equipment Rooms for Boilers and Chillers

A boiler room, if required by the application, will house one or more boilers and required support equipment. For a hot water boiler system, space for pumps must be provided. For a steam boiler system, the space requirements are significantly greater, requiring support space for the feedwater and/or deaerators and water treatment systems.

Chiller rooms must be separate from boiler rooms and require space for the chiller(s), chilled water pump(s), condenser water pumps, and water treatment systems. ASHRAE Standard 15, adopted by many code authorities, classifies a chiller room as a "refrigeration machinery room" if the amount of refrigerant in the largest chiller exceeds certain specified levels. *If so classified, the room must be constructed to higher standards and a detailed code review for these spaces is critical.*

The space required for a boiler room and chiller room can vary significantly due to the number and arrangement of boilers and chillers. A reliable primary system will include at least two boilers and/or chillers to provide standby capacity in the event of equipment failure. Hospitals, research facilities, etc. will have at least three boilers and chillers (each sized for 50% of the load) so that the failure of one unit will not impact the ability to meet peak loads.

As a general guide, the following are the *minimum* space allocations for primary HVAC equipment rooms:

Hot Water Boiler Room: 0.7 % of the gross building area

Steam Boiler Room: 1.0 % of the gross building area

Chiller Room: 1.5-1.7 % of the gross building area

Room Height: 14'-18' clear below the bottom or the roof or floor above structural elements

On construction documents, show tube pull/clean space requirements so that piping, raceway, or other equipment does not intrude into these spaces.

The designers must provide flood control measures in all primary HVAC equipment rooms. All floor openings must be protected with 2- to 4-inch-high curbs to dam loose water on the floor. Floor drains must be provided adjacent to pumps, chillers, boilers, cooling towers, etc., with floors sloped to each drain.

Spill control is also required around areas containing water treatment chemicals. The most cost-effective approach is for the HVAC engineer to specify the use of double walled containers, with spill lips, for all chemicals. If this is not possible, a 6" high curb is required to contain the volume of the largest storage container. A 6"-12" deep "pump out" recess is also required to facilitate cleanup of a chemical spill.

Ceiling Cavity Space

Sufficient ceiling space for HVAC air distribution (ductwork); HVAC, plumbing, and fire protection piping; and electrical raceway is required. These space requirements are in addition to the minimum space required for the ceiling itself and recessed lighting fixtures.

Table 5-2 reflects typical minimum ceiling space requirements for commercial and institutional building when air secondary systems are used.

Table 5-2. Ceiling Space Requirements for Maintainability

Lighting:	8"	(6" for fixture plus 2" to lift and move fixture above the ceiling.)
Ductwork:	24"	(Ducts with high aspect, ratio width to height, increase installation costs.)
Piping and Electrical Raceway:	12"	(May increase to incorporate fire protection and/or roof drainage systems)
Total:	44"	

Thus, the minimum floor-to-floor height in a multistory building is 44" + floor-to-ceiling height + structure depth.

When water type secondary systems are used, the ductwork space requirement can be eliminated in single story buildings and reduced to only 12"-18" in multistory buildings.

The space requirements listed above must be increased for special occupancies such as laboratories, hospitals, kitchens, etc. because of the much higher "density" of distribution systems in these types of facilities.

ELECTRICAL

Adequate space and access for maintenance of electrical equipment must be provided. While the NFPA 70, *National Electrical Code*, sets strict requirements for service clearances, these service clearances may be impeded by the installation of HVAC piping, ductwork, and equipment or other space use (including, commonly, storage use by the owner). Careful consideration of these clearance requirements by the design team is a must.

Appendix A

Substructure Preventative and Predictive Maintenance Procedures

The facility substructure includes footings, foundations, slab-on-grade, and basement construction.

1.1.1.0—Settlement
Inspect annually for settlement cracks, spalling, etc. Retain Professional Engineer for more extensive investigation if indicated by the extent of the problem(s). Ensure that organic matter is cleared away at least 8" below floor level and that grade slopes away with at least 5% slope.

1.2.0.0—Under-slab Vapor Barrier
Failure of under-slab vapor barrier can cause slab damage due to moisture migration. Inspect annually for bubble or other signs of this problem. Retain Professional Engineer for more extensive investigation if indicated by the extent of the problem(s).

1.3.0.0—Moisture Penetration
Inspect annually for evidence of moisture penetration to indoors—flaking or bubbling paint, soft or discolored GWB, etc. Check for wall cracks. Retain Professional Engineer for more extensive investigation if indicated by the extent of the problem(s).

Appendix B

Shell Preventative and Predictive Maintenance Procedures

The facility "shell" consists of the superstructure including walls, windows, exterior doors, roofs, etc.

2.1.1.0—Floor Construction
Inspect building floor construction every 5 years. Repair as necessary.

2.1.1.1—Deck Waterproofing
Exterior deck waterproofing systems must be inspected annually, as follows:

1. Inspect deck surface waterproofing for signs of deterioration and for damage caused from deck surface expansion/contraction of the substructure which may have visible plywood seams.

2. Inspect deck to wall sheet metal flashing and counter flashing for deterioration and damage.

3. Inspect deck to door threshold sheet metal flashing for deterioration and for damage caused from wear and tear.

4. Inspect perimeter edge sheet metal flashing for lap separation and for deck surface waterproofing separation at the edge metal juncture.

5. Inspect railing post deck penetrations for open fractures.

6. Inspect sheet metal scupper drains and through deck drains for open fractures and for wear and tear.

Make repairs or replace elements as necessary.

2.1.1.3—Gutters and Downspouts

Inspect gutters and downspouts every 6 months and repair, as follows:

Gutters

1. Check for blockage, clogging, corrosion, loose or cracked joints, plant growth, and leaks. Remove debris, clear blockages or clogs, and remove plant growth.

2. Wash out gutter with garden hose and check for leaks. Any leak in a gutter may be patched (if from puncture or corrosion) or soldered (if located at a joint).

3. Check for proper pitch of the gutter for adequate drainage. If water does not drain properly, re-hang gutter for proper pitch.

4. Check for corroded, broken and loose fasteners. Repair/replace as required.

5. Check each gutter/downspout connection. Each should have an aluminum or stainless steel screen to prevent large debris from entering. Install screen if missing or corroded.

Downspouts

1. Ensure all gutters have downspouts that direct the water to extensions, splash-backs, or site drainage systems that direct water at least three feet away from the foundation and clear any blockage from downspout outlet. If downspouts connect directly to site drainage piping, confirm that there are no blockages at the connection (see Maintenance Procedure 7.2.2.4).

2. Check downspouts for water flow and leaks. Check all joints in the downspouts, ensuring no breaks. Any leak may be patched (if from puncture or corrosion) or soldered (if located at a joint).

3. Check hanging brackets of downspouts, ensuring they are tight. Repair/replace as needed.

4. Check that lateral downspout section have at least 30° pitch, without sags. Remove and re-hang as needed.

2.2.1.0—Concrete Panels

Inspect concrete panels annually.

1. Check surface appearance, connections, etc.. Any signs of deterioration (cracking, spalding, movement, etc.) require further investigation by a Professional Engineer.

2. Clean panels every 5 years via power washing with non-acid cleaning solution. Inspect surface for cracks, spalling, loss of surface aggregate (if exposed), exposed reinforcement, rust streaks at connection points, etc. Examine panels for movement or "racking," which could indicate a connection or structural failure. Reseal panels immediately after cleaning with clear concrete sealer.

3. Inspect and repair/replace caulking in accordance with Maintenance Procedure 2.2.1.8.

If staining from window cleaning is found, report to facilities director so the window cleaning process can be changed.

2.2.1.1—Brick Masonry

Inspect brick masonry annually and repair as required, as follows:

WHAT TO LOOK FOR	REMEDIAL ACTIONS
Cracked, loose, or spalled bricks	Cracked, loose, or spalled bricks must be replaced.
	The mortar that surrounds the affected units must be cut out carefully to avoid damaging adjacent brickwork. For ease of removal, the brick to be removed can be broken. Once the units are removed, all of the surrounding mortar should be carefully chiseled out, and all dust and debris should be swept out with a brush. If the units are located in the exterior wythe of a drainage wall (which is the typical case), care must be exercised to prevent debris from falling into the air space, which could block weeps and interfere with moisture drainage. The brick surfaces in the wall should be dampened before new units are placed, but the

masonry should absorb all surface moisture to ensure a good bond. The appropriate surfaces of the surrounding brickwork and the replacement brick should be buttered with mortar. The replacement brick should be centered in the opening and pressed into position. The excess mortar should be removed with a trowel. Pointing around the replacement brick will help to ensure full head and bed joints. When the mortar becomes "thumbprint" hard, the joints should be tooled to match the original profile.

Bulging bricks or wall out-of-plumb	Typically, this results from missing or failed brick ties and anchors. Retrofit ties/anchors must be installed.
	Installation of most retrofit anchors involves drilling small holes in the masonry, usually in a mortar joint, through which the anchors are attached to the substrate. Generally, mechanical expansion, helical screws, grout- or epoxy-adhesive systems are used. Because the installation methods and limitations of each product are unique, consultation with the manufacturer is essential to assure proper application, detailing, installation, inspection, and performance.
Hairline cracks in mortar	Cracks in brickwork that are more than a few millimeters in width or that are suspected to have been caused by settlement or other structural problems (for example, cracks that continue through multiple brick units and mortar joints, or follow a stepped or diagonal pattern along mortar joint) require professional investigation to determine the cause and appropriate method of repair.
	If the mortar joints develop small "hairline" cracks, surface grouting should be done to fill them. A recommended grout mixture is 1 part portland cement, 1/3 part hydrated lime and 1-1/3 parts fine sand (passing a No. 30 sieve). The joints to be grouted should be dampened. To ensure good bond, the brickwork must absorb all surface water. The grout mixture should be applied to the joints with

a stiff fiber brush to force the grout into the cracks. Two coats are usually required to effectively reduce moisture penetration. Tooling the joints after the grout application may help compact and force the grout into the cracks.

Deteriorated mortar joints	Repointing is the process of removing damaged or deteriorated mortar to a uniform depth and placing new mortar in the joint.

Deteriorated mortar should be removed by means of a toothing chisel or a special pointer's grinder to a uniform depth that is twice the joint width or until sound mortar is reached. Care must be taken not to damage the brick edges. Remove all dust and debris from the joint by brushing, blowing with air or rinsing with water.

Type N, O and K mortar are recommended for repointing. The repointing mortar should be pre-hydrated to reduce excessive shrinkage. The proper pre-hydration process is as follows: All dry ingredients should be thoroughly mixed. Only enough clean water should be added to the dry mix to produce a damp consistency which will retain its shape when formed into a ball. The mortar should be mixed to this dampened condition 1 to 1-1/2 hr before adding water for placement.

The joints to be repointed should be dampened, but to ensure a good bond, the brickwork must absorb all surface water before repointing mortar is placed. Water should be added to the prehydrated mortar to bring it to a workable consistency (somewhat drier than conventional mortar). The mortar should be packed tightly into the joints in thin layers (1/4 in. maximum). The joints should be tooled to match the original profile after the last layer of mortar is "thumbprint" hard. As it may be difficult to determine which joints allow moisture to penetrate, it is advisable to repoint all mortar

	joints in the affected wall area.
Missing or clogged weepholes	Weeps should be inspected to ensure that they are open and appropriately spaced so that moisture within the walls is able to escape to the exterior. If weeps are clogged, they can be cleaned out by probing with a thin dowel or stiff wire. If the weeps were not properly spaced, drilling new weeps may be necessary. Since weeps are placed directly above flashing, care must be exercised to not damage the flashing when probing or drilling. The use of a stopper to limit the depth of penetration of the probe or drill bit may be effective in reducing the possibility of damaging the flashing where it turns up inside of the brick wythe. In order to properly drain any water collected on the flashing, weeps are required immediately above the flashing at all locations. An open head joint, formed by leaving mortar out of a joint, is the recommended type of weep. Open head joint weeps should be at least 2 in. high. Weep openings are permitted by most building codes to have a minimum diameter of 3/16 in. The practice of placing weeps in one or more courses of brick above the flashing can cause a backup of water and is not recommended. Non-corrosive metal, mesh or plastic screens can be installed in open head joint weeps if desired. Spacing of open head joint weeps is recommended at no more than 24″ on center. Spacing of wick and tube weeps is recommended at no more than 16 in. on center. Weep spacing is permitted by most building codes at up to 33 in. on center. Wicks should be at least 16 in. long and extend through the brick, into the air space and along the back of the brick.
Efflorescence or stains	Efflorescence is water-soluble and easily removed by natural weathering or by scrubbing with a brush and water. Proprietary cleaners formulated specifically for use on brickwork are effective in removing stubborn efflorescence. *Do not use acid cleaning.*

	Old stains caused by improper cleaning are not water-soluble, but can be removed by proprietary cleaners.
Water penetration	The two primary causes of water penetration are poor mortar joints and/or failed or clogged through flashing.
	New through flashing can be installed. *This is a difficult procedure of removing brick, installing flashing, and replacing the units removed.*
	To install continuous flashing in existing walls, alternate sections of masonry in 2 to 5 ft lengths should be removed. The flashing is installed in these sections and the masonry replaced. Alternately, temporary braces can be installed as longer sections of brickwork are removed. The flashing can then be placed in these sections. The lengths of flashing should be lapped a minimum of 6 in. and be completely sealed to function properly. The opening is then filled as described for brick replacement above. The replaced masonry should be properly cured (5 to 7 days) before the intermediate masonry sections or supports are removed.
	An alternative approach is to cover the wall to reduce water loading. EFIS, metal panels, or acrylic elastomeric coating can be used for this purpose. Do not apply clear sealant; it is ineffective.
Brick or stone coping, cap, or sill problems	Check for cracked elements, loose or open joints, hairline mortar cracks, etc. and repair as described above.
Plant growth, algae	To remove ivy and similar plants, the vines should be carefully cut away from the wall...never pulled from the wall as this could damage the brickwork. After cutting, the shoots will remain. These suckers should be left in the wall until they dry up and shrivel, which usually takes 2 to 3 weeks. Care should be taken not to allow the suckers to rot as this could make them difficult to remove. Once

the shoots dry, the wall should be dampened and scrubbed with a stiff fiber brush and water. Laundry detergent or weed killer may be added to the water in small concentrations to aid in the removal of the shoots. If these additives are used, the wall must be thoroughly rinsed with clean water before and after scrubbing.

Remove algae growth by light pressure washing—make sure not to damage mortar.

2.2.1.2—Exterior Insulation and Finishing Systems (EFIS)
Clean and inspect EFIS annually.
Clean EFIS as follows:

1. Remove dirt, mildew, and algae with the use of general cleaning compounds, followed by a mild pressurized water clean and rinse. This can usually be accomplished with a garden hose and an ordinary nozzle. No acidic cleaners are recommended. (If power washing by a professional, keep pressure under 600 psi and use cold, unheated water. Hold sprayer with a fan-tip nozzle at a 45° angle from the wall (not perpendicular) and keep spray tip at least 2' 0" from surface.) Also, do not use abrasive hard-bristle brushes.

2. Paint or other stains may require special treatment to remove. Depending on what type of paint used, it may or may not be possible to remove. Try cleaning as described above. Do not use solvents or solvent-based cleaners such as paint thinner or mineral spirits. If the paint cannot be removed with proper cleaning techniques, it will be necessary to refinish or paint over it.

3. Asphalt or tar stains cannot be chemically removed. These stains must be chipped off and the affected area must be refinished.

4. Graffiti may be removed by lightly scrubbing with a soft-bristle brush and a solution of 1 gallon warm water and 1 cup of liquid soap. This may not work on oil-based spray paints, and the surface may need to be refinished.

After cleaning, carefully inspect around doors and windows for cracks in the exterior stucco and sealant. Add and/or replace sealant as needed (see Maintenance Procedure 2.2.1.8).

2.2.1.3—Vinyl and Aluminum Siding

Inspect vinyl and aluminum siding annually and after major storms. Replace any missing siding sections. Bent or curved sections that won't go back into place need to be replaced.

Vinyl must be cleaned annually to eliminate dirt, dust, grass stains, mold or mildew. For heavier stains, vinyl siding can be cleaned with a soft cloth, or soft bristled toothbrush (if the surface is textured). Some of the more common cleaners that can be used on vinyl include: Fantastic, Murphy's Oil Soap, Lestoil, Windex, and Soft Scrub.

Since vinyl and its accessories will melt when exposed to significant heat, clear mulch and dry leaves away from the vinyl.

Aluminum finishes will fade after 5 years, requiring painting then and every 5-7 years thereafter.

Aluminum may need to be cleaned to remove dirt, grass, or mildew. (Pressure washing aluminum is possible but should first be done with plain water on low pressure. If low pressure doesn't seem to be making the area clean do a higher power test wash on an area that is less visible. The test wash is important because some pressure washers create enough pressure to dent or deform the aluminum panels.) It is best to use a biodegradable detergent and hose it off with a light spray.

Aluminum should be inspected for dents or scratches. Damaged panels should either be painted (minor damage) or replaced (serious damage).

2.2.1.4—Wood Siding

Inspect wood siding every 6 months for the following conditions:

- Damaged wood (holes, cracks, warped boards, wet rot, dry rot)
- Damaged or failing paint
- Caulking failure (see Maintenance Procedure 2.2.1.8.)
- Insect damage

Make repairs immediately. More importantly, as damage is repaired, identify what caused the damage and correct that problem, also. For example, if wood siding is damaged by leaky gutters or pour drainage from downspouts, correct these problems. Likewise, failing flashing above doors or windows may be a problem.

If paint problems such as blistering, peeling, alligatoring, wrinkling, or chalking are confined to a small area, you can touch them up. If they are more extensive, repainting of the entire facility is required (see Maintenance Procedure 2.2.1.12).

Test for wet rot or dry rot with a screwdriver blade or ice pick and replace all rotted wood. Where only the lower parts of the wood are damaged, these can be cut away, a new z-flashing installed, and new siding installed. Prime *all edges* and back of new wood before installation. Structural wood that is damaged must be replaced with treated wood.

Insect damage indicates the need for further investigation and treatment by a pest control contractor.

2.2.1.5—Cement-asbestos Board

Inspect cement-asbestos board or panel every two years for breaks, cracks, holes, etc. A damaged board or panel cannot be repaired. It can only be replaced with like material or covered with another material.

The asbestos in cement-asbestos ("transite") is non-friable and must be maintained so throughout any repair or replacement activity. *The board must not be broken* as this will release asbestos fibers into the air and, thus, great care is required. After removal, place the board in a sealed plastic back for landfill disposal.

2.2.1.6—Stucco

Stucco must be inspected annually.

The basic conditions associated with the deterioration of stucco are:

* Biological growth (anything from molds to vines and plants)
* Blistering (looks like bubbles beneath the stucco)
* Cracking
* Delamination (the finish, brown, and scratch coats detach from one another)
* Detachment (the stucco actually detaches from the wall beneath)
* Disaggregation (loss of binder)
* Flaking (thin, small layers of the finish coat loosen and flake off)
* Loss (sometimes sheets of stucco detach and fall exposing the substrate beneath)

Most often, water causes these conditions. Repeated freezing and thawing causes internal deterioration, detachment of stucco from the substrate, and delamination of the stucco's layers. Excessive water infiltration (through cracks or loss, for example) leads to a high moisture content which presents a hospitable environment for biological growth —not only an eyesore, but root systems can also lead to deterioration.

Cracking in stucco can generally be attributed to building settlement or water damage. Rotting of wood or corrosion of metal lath as a result of excessive water infiltration can lead to detachment and loss of the stucco.

Repairs may include a combination of approaches, including crack filling, patching, and sometimes injections and grouting. It is critical to use compatible materials with respect to mechanical as well as aesthetic properties. The resulting finish should blend visually with the existing stucco in both texture and color. Most importantly, repair materials used should be weaker and more permeable than surrounding stucco.

2.2.1.7—Concrete Masonry

Inspect concrete masonry unit (CMU) walls annually and repair as required, as follows:

WHAT TO LOOK FOR	REMEDIAL ACTIONS
Cracked or loose units	Cracked or loose units must be replaced.
	The mortar that surrounds the affected units must be cut out carefully to avoid damaging adjacent brickwork. For ease of removal, the units to be removed can be broken. Once the units are removed, all of the surrounding mortar should be carefully chiseled out, and all dust and debris should be swept out with a brush. If the units are located in the exterior wythe of a drainage wall (which is the typical case), care must be exercised to prevent debris from falling into the air space, which could block weeps and interfere with moisture drainage. The CMU surfaces in the wall should be dampened before new units are placed, but the masonry should absorb all surface moisture to ensure a good bond. The appropriate surfaces of the surrounding CMU and the replacement unit should be buttered with mortar. The replacement unit should be centered in

the opening and pressed into position. The excess mortar should be removed with a trowel. Pointing around the replacement unit will help to ensure full head and bed joints. When the mortar becomes "thumbprint" hard, the joints should be tooled to match the original profile.

Cracks in mortar joints	If cracks are fine, they should not be widened. It may be adequate to paint over them with portland cement paint. Otherwise, a mortar of one part portland cement and one part sand that passes a No. 30 or No. 50 screen, depending on the size of the crack, should be worked in.
	Larger cracks can be readily repaired by tuckpointing (see "Deteriorated Mortar Joints" below)
	The other way to repair cracks is to widen and undercut them with saw or chisel so that they are about one half inch wide at the surface, 1/2 to 3/4 inch wide at the back, and about one half inch deep. All loose dust should be brushed free. The notch should then be filled either with a dry-pack mortar or a non-sagging epoxy resin, If mortar is used, the surrounding block should be dampened thoroughly to minimize absorption of water from the mortar. A mortar of one part portland cement and two parts masonry sand should be mixed to a stiff consistency and packed in. If epoxy resin is used, it should be one that produces a gel of grease-like consistency when first mixed. When the mortar gets "thumbprint" hard, tool the joint to match existing.
Deteriorated mortar joints	Tuckpointing is the process of removing damaged or deteriorated mortar to a uniform depth and placing new mortar in the joint.
	The deteriorated mortar should be removed, by means of a toothing chisel or a special pointer's grinder, to a uniform depth that is twice the joint width or until sound mortar is reached. Care must be taken not to damage the brick edges. Remove all dust and debris from the joint by brushing, blowing with air or rinsing with water.

Type N, O and K mortar are generally recommended for repointing, as mortars with higher cement contents may be too strong for proper performance. The repointing mortar should be prehydrated to reduce excessive shrinkage. The proper prehydration process is as follows: All dry ingredients should be thoroughly mixed. Only enough clean water should be added to the dry mix to produce a damp consistency which will retain its shape when formed into a ball. The mortar should be mixed to this dampened condition 1 to 1-1/2 hr before adding water for placement.

The joints to be repointed should be dampened, but to ensure a good bond, the brickwork must absorb all surface water before repointing mortar is placed. Water should be added to the prehydrated mortar to bring it to a workable consistency (somewhat drier than conventional mortar). The mortar should be packed tightly into the joints in thin layers (1/4 in. maximum). The joints should be tooled to match the original profile after the last layer of mortar is "thumbprint" hard. As it may be difficult to determine which joints allow moisture to penetrate, it is advisable to repoint all mortar joints in the affected wall area.

Missing or clogged weepholes	Weeps should be inspected to ensure that they are open and appropriately spaced so that moisture within the walls is able to escape to the exterior. If weeps are clogged, they can be cleaned out by probing with a thin dowel or stiff wire. If the weeps were not properly spaced, drilling new weeps may be necessary.

Since weeps are placed directly above flashing, care must be exercised to not damage the flashing when probing or drilling. The use of a "stop" to limit the depth of penetration of the probe or drill bit may be effective in reducing the possibility of damaging the flashing where it turns up inside of the brick wythe.

In order to properly drain any water collected on the flashing, weeps are required immediately above the flashing at all locations. An open head joint, formed by leaving mortar out of a joint, is the recommended type of weep. Open head joint weeps should be at least 2 in. high. Weep openings are permitted by most building codes to have a minimum diameter of 3/16 in. The practice of placing weeps in one or more courses of brick above the flashing can cause a backup of water and is not recommended. Non-corrosive metal, mesh or plastic screens can be installed in open head joint weeps if desired. Spacing of open head joint weeps is recommended at no more than 24″ on center. Spacing of wick and tube weeps is recommended at no more than 16 in. on center. Weep spacing is permitted by most building codes at up to 33 in. on center. Wicks should be at least 16 in. long and extend through the brick, into the air space and along the back of the brick.

Water penetration	The two primary causes of water penetration are poor mortar joints and/or failed or clogged through flashing.

New through flashing can be installed. This is a difficult procedure of removing CMUs, installing flashing and replacing the units may be required. To install continuous flashing in existing walls, alternate sections of masonry in 2 to 5 ft. lengths should be removed. The flashing is installed in these sections and the masonry replaced. Alternately, temporary braces can be installed as longer sections of brickwork are removed. The flashing can then be placed in these sections. The lengths of flashing should be lapped a minimum of 6 in. and be completely sealed to function properly. The opening is then filled as described for CMU replacement above. The replaced masonry should be properly cured (5 to 7 days) before the intermediate masonry sections or supports are removed.

An alternative approach is to cover the wall. EFIS, metal panels, or acrylic elastomeric coating can be used for this purpose. Do not apply clear sealant...it is not effective.

Masonry or stone coping, cap, or sill problems	Check for cracked elements, loose or open joints, hairline mortar cracks, etc. and repair as described above.
Plant growth, algae	To effectively remove ivy and similar plants, the vines should be carefully cut away from the wall. The vines should never be pulled from the wall as this could damage the brickwork. After cutting, the shoots will remain. These suckers should be left in the wall until they dry up and shrivel. This usually takes 2 to 3 weeks. Care should be taken not to allow the suckers to rot as this could make them difficult to remove. Once the shoots dry, the wall should be dampened and scrubbed with a stiff fiber brush and water. Laundry detergent or weed killer may be added to the water in small concentrations to aid in the removal of the shoots. If these additives are used, the wall must be thoroughly rinsed with clean water before and after scrubbing. Remove algae growth by light pressure washing... make sure not to damage mortar joints.

2.2.1.8—Caulked Joints

Inspect all caulked joints every 3 years for evidence of deterioration or failure. Alligatoring and limited separation (5% or less) are signs that the caulking is aging and may be nearing the end of its life. Significant separation (30% or more) and loose or lost caulking are signs of caulking failure. In the first case, repairs may extend the life of the caulk by 2-4 years. In the second case, the only option is replacement.

Repair
1. Clean and soften existing caulk by wiping with cloth in appropriate solvent.

2. Apply thin coating of new caulk over existing, being careful to fill splits and separated sections to a dept of at least 1/8."

3. Apply only flexible caulk, as follows:

Vertical Surfaces: Silicone or urethane (polyurethane). ASTM C920, Type S, Grade NS, Class 25, Use M, A, or O, as applicable.

Horizontal Surfaces: Urethane (polyurethane). ASTM C920, Type S or M, Grade P, Class 25, Use T. Grade NS, Use T, in areas with slopes exceeding 1 percent.

Note: Silicone caulk will adhere to urethane caulk, but urethanes will not adhere to silicone. If in doubt about existing caulk, use silicone for repairs.

4.   Do not use caulk that has been stored too long…check shelf life.

5.   Apply caulk at the proper temperature, defined by the manufacturer on the tube. Do not apply when the drying period will be humid (above 70% RH) or rainy.

Replacement

1.   Remove existing caulking by scraping or grinding, being careful not to damage the adjacent materials.

2.   Clean concrete, masonry, and similar porous joint substrate surfaces by brushing, grinding, blast cleaning, mechanical abrading, acid washing, or a combination of these methods to produce a clean, sound substrate capable of developing optimum bond with joint sealers. Clean metal, glass, porcelain enamel, glazed surfaces of ceramic tile and other nonporous surfaces by chemical cleaners or other means which are not harmful to substrates or leave residues capable of interfering with adhesion of joint sealers. Do a final wipe with the appropriate solvent. Prime or seal joint surfaces where recommended by the sealant manufacturer.

3.   If backer rods are in place, avoid damaging them. Check backer rods to determine if they are in good condition and will remain in place during re-caulking. If backer rods are missing or damaged, install new ones. Backer rods should have a diameter 30% greater than the width of the joint being caulked.

4.   Test new caulk at an inconspicuous spot to ensure that it will adhere to the existing surfaces.

5.   Apply only flexible caulk, as follows:

     Vertical Surfaces: Silicone or urethane (polyurethane). ASTM C920, Type S, Grade NS, Class 25, Use M, A, or O, as applicable.

     Horizontal Surfaces: Urethane (polyurethane). ASTM C920, Type S or M, Grade P, Class 25, Use T. Grade NS, Use T, in areas with slopes exceeding 1 percent.

6.   Do not use caulk that has been stored too long…check shelf life.

7.   Depth of caulk joint should not exceed the width of the joint and preferably should be only half the width. Where the dept of the joint is shallow and backer rod cannot be used, utilized "bond-breaker" tape to avoid 3-side adhesion.

8.   Apply the caulk at the proper temperature, defined by the manufacturer on the tube. Do not apply when the drying period will be humid (above 70% RH) or rainy.

2.2.1.9—Exterior Wall Flashings
Inspect exterior wall flashings, counter-flashings, and drips every 5 years. Repair or replace as necessary.

2.2.1.10—Painted Steel Lintels
Inspect painted steel lintels annually. Maintain paint in accordance with Maintenance Procedure 2.2.1.11. Repair or replace caulked joints in accordance with Maintenance Procedure 2.2.1.8.

2.2.1.11—Walls Exposed to Sun or Salt Spray
Inspect paint on the south- and west-facing walls or walls exposed to salt spray every 6 months. Inspect all other painted surfaces annually.

WHAT TO LOOK FOR	PROBABLE CAUSE	REMEDIAL ACTIONS
Alligatoring (Patterned cracking in the surface of the paint film resembling the regular scales of an alligator.)	Application of an extremely hard, rigid coating, like an alkyd enamel, over a more flexible coating, like a latex primer. Application of a top coat before the undercoat is dry. Natural aging of oil-based paints as temperatures fluctuate. The constant expansion and contraction results in a loss of paint film elasticity.	Old paint should be completely removed by scraping and sanding the surface; a heat gun can be used to speed work on large surfaces, but take care to avoid igniting paint or substrate. The surface should be primed with a high quality latex or oil-based primer, then painted with exterior latex paint.
Blistering (Formation of bubbles resulting from localized loss of adhesion and lifting of the paint film from the underlying surface.)	Painting in direct sunlight or on a surface that is too warm, especially when applying a dark-colored solvent-based coating applying an oil-based or alkyd paint over a damp or wet surface. Excess humidity or other moisture escaping from inside through the exterior walls (less likely with latex paints, which allow water vapor to escape without affecting the paint film). Exposure of a latex paint film to excessive moisture in the form of dew, high humidity or rain shortly after the paint has dried,	First, determine whether or not the blisters go all the way down to the substrate. If so, the problem may be due to moisture coming from inside. Take steps to remove the source of moisture, if possible. Repair loose caulk and consider installing vents or exhaust fans. If the building has wood siding, install siding vents in areas where blistering has occurred. Remove blisters by scraping and sanding, prime any areas where bare wood shows, and repaint.

	especially if there was inadequate surface preparation and/or a lower quality paint was applied	If the blisters do not go all the way to the substrate, the problem is probably not related to moisture coming from behind. Rather, the blisters are likely from painting a warm surface in direct sunlight or exposing the paint film to excessive moisture. In any case, sand, scrape and then prime any exposed bare wood. Coat with acrylic latex exterior paint.
Cracking/flaking (splitting of a dry paint film through at least one coat, which will lead to complete failure of the paint. Early on, the problem appears as hairline cracks; later, flaking of paint chips occurs.)	Use of a lower quality paint that has inadequate adhesion and flexibility. Over-thinning the paint or spreading it too thin. Poor surface preparation, especially when the paint is applied to bare wood without priming. Painting under cool or windy conditions that make latex paint dry too fast.	It may be possible to correct cracking that does not go down to the substrate by removing the loose or flaking paint with a scraper or wire brush, sanding to feather the edges, priming any bare spots and repainting. If the cracking goes down to the substrate, remove all of the paint by scraping, sanding and/or use of a heat gun; then prime and repaint with a quality exterior latex paint.
Nailhead rusting	Non-galvanized iron nails have begun to rust, causing bleed-through to the top coat. Non-galvanized iron nails have not been countersunk and filled over.	When repainting exteriors where nailhead rusting has occurred, wash off rust stains, sand the nailheads, then countersink the nailheads, caulk them with water-based all-acrylic or siliconized acrylic caulk. Each

	Galvanized nailheads have begun to rust after sanding or excessive weathering.	nailhead area should be spot primed, then painted with a quality latex coating.
Loss of paint adhesion	Incompatible paints such as use of water-based latex paint over more than three or four coats of old alkyd or oil-based paint may cause the old paint to "lift off" the substrate.	Repaint using another coat of alkyd or oil-based paint. Or completely remove the existing paint and prepare the surface cleaning, sanding and spot-priming where necessary, before repainting with latex exterior paint.
Peeling	Swelling of wood due to seepage or penetration of rain, humidity and other forms of moisture into the building through un-caulked joints; deteriorated caulk; leaking roof or other areas excess humidity or other moisture escaping from within through the exterior walls (less likely with latex paints, which allow water vapor to escape without affecting the paint film); inadequate surface preparation; use of a lower quality paint that has inadequate adhesion and flexibility characteristics; applying latex paint under conditions that hinder good film formation, e.g., on a very hot or	Eliminate the cause by doing necessary repairs and maintenance: replace caulking; repair roof; clean gutters and downspouts; cut heavy vegetation away from the building. Remove all loose paint with a scraper or wire brush, sand rough surfaces and prime any bare wood. Repaint with acrylic latex paint to provide adhesion and allow water vapor to escape without harming the coating.

	cold day; in windy weather applying an oil-based paint over a damp or wet surface.	
Painted vinyl side warp or buckling	Vinyl siding was repainted with a darker color paint than the original color. Dark paint tends to absorb the beat of the sun, transferring it to the substrate. When vinyl siding expands dramatically, it is not able to contract to its original dimensions.	Given the option, do not paint vinyl! If it must be painted, paint in a shade no darker than the original. Whites, off-whites, pastels and other light colors are good choices. Acrylic latex paint is the best type of paint to use on vinyl siding, because the flexibility of the paint film enables it to withstand the stress of expansion and contraction cycles caused by outdoor temperature changes. Siding that has warped or buckled should be assessed to determine the best remedy. The siding may have to be replaced.
Wrinkling (Rough, crinkled paint surface occurring when paint forms a "skin.")	Paint applied too thickly (more likely when using alkyd or oil-based paints). Painting a hot surface or in very hot weather. Exposure of uncured paint to rain, dew, fog or high humidity levels. Applying top coat of paint to insufficiently dried first coat. Painting over contaminated surface (e.g., dirt).	Scrape or sand substrate to remove wrinkled coating. Repaint, applying an even coat of exterior paint. Make sure the first coat or primer is dry before applying the top coat. Apply paints at the manufacturer's recommended spread rate (two coats at the recommended spread rate are better than one thick coat). When painting during extremely hot, cool or damp weather, allow extra time for the paint to dry completely.

2.2.1.12—Metal Building Wall Panels
Inspect metal building wall panels every 5 years.

WHAT TO LOOK FOR	PROBABLE CAUSE	REMEDIAL ACTIONS
Corrosion	General rusting indicates that the paint finish and zinc coating is no longer protecting the underlying steel.	Thoroughly clean panels, remove rust by sandblasting or sanding, apply zinc primer and final coat(s) of acrylic latex paint in accordance with Maintenance Procedure 2.2.1.11.
	Scratches and dings.	Clean, sand, and touch-up areas with manufacturer provided touch-up materials.
	Localized rusting, particularly at fasteners, indicates galvanic corrosion.	Localized rusting requires fastener replacement with rubber/EPDM grommets and washers to eliminate contact between roof and fastener. Remove rust by sanding, priming with zinc chromate or other rusty metal primer, and application of a finish coat to match existing before installing fasteners.
	Rusting due to ground contact	Keep ground litter away from panels...bottom of panel should be at least 8" above grade. If necessary, remove dirt to provide clearance and at least 5% slope away from building wall.

Splits, holes, or breaks	Mechanical damage	Replace damaged panel(s).
Water leaks (indicated by interior water stains or sagging roof insulation)	Neoprene washers are widely used at fasteners, but dry and crack over time.	Fastener leaks due to washer failure can be repaired by replacing washers with EPDM type in lieu of original neoprene washers.
Moisture infiltration	Flashing problems, including rust, separation, etc.	Replace deteriorated or damaged flashing materials.

2.2.2.0—Steel Windows

Inspect steel windows every two years. Determine condition of all three parts (sash, frame, and sub-frame) and perform repairs essential to continued use. This evaluation and repairs should include the following:

WHAT TO LOOK FOR	REMEDIAL ACTIONS
Corrosion	Use a sharp probe or tool, such as an ice pick, to determine the extent of corrosion in the metal: 1. If the rusting is merely a surface accumulation or flaking, then the corrosion is light. 2. If the rusting has penetrated the metal (indicated by a bubbling texture), but has not caused any structural damage, then the corrosion is medium. 3. If the rust has penetrated deep into the metal, the corrosion is heavy. Heavy corrosion generally results in some form of structural damage, through delamination, to the metal section, making repair difficult and expensive and the window should be replaced. In any case, since moisture is the primary cause of corrosion, it is essential that excess moisture be eliminated before any repair work is undertaken.

To maintain windows, remove light and medium rust via manual abrasion (wire brush and aluminum oxide sandpaper) and mechanical abrasion (such as an electric drill with a wire brush or a rotary whip attachment). Rust can also be removed by using a number of commercially prepared anticorrosive acid compounds. If chemicals are used, any chemical residue should be wiped off with damp cloths, then dried immediately with Industrial blow-dryers. Do not use running water to remove chemical residue.

Removing rust will remove most flaking paint as well. Remaining loose or flaking paint can be removed with a chemical paint remover or with a pneumatic needle scaler or gun. (Well-bonded paint may serve to protect the metal further from corrosion and need not be removed. The paint edges should be feathered by sanding to give a good surface for repainting.)

Once metal has been cleaned of all corrosion, small holes and uneven areas resulting from rusting should be filled with a patching material of steel fibers and an epoxy binder and sanded smooth to eliminate pockets where water can accumulate.

Bare metal should then be wiped with a cleaning solvent such as denatured alcohol, and dried immediately in preparation for the application of an anticorrosive primer (oil-alkyd based paint rich in zinc or zinc chromate), applied immediately after cleaning.

Broken, missing, or damaged glass	Remove broken glass and save all clips, glazing beads, and other fasteners that hold the glass to the sash, although replacements for these parts are still available.
	Double glazing where the air space has a "cloudy" or "milky" film has failed and must be replaced.
	Clean mullions thoroughly and install glass with glazing compound specifically formulated for steel windows.
Failed glazing compound	Remove broken, dried, or separated glazing compound. If necessary, replace glazing beads, and other fasteners that hold the glass to the sash.

	Clean mullions thoroughly and install glass with glazing compound specifically formulated for steel windows.
Hinges	Use a cleaning solvent and fine bronze wool to clean the hinges (generally brass or bronze). The hinges should then be lubricated with a non-greasy lubricant specially formulated for metals and with an anticorrosive agent. These lubricants are in a spray form and should be used once or twice a year on frequently opened windows.
Weather stripping	Install weatherstripping where the operable portion of the sash and the fixed frame come together to reduce perimeter air infiltration. Where possible, use a spring-metal type weatherstripping with an integral friction fit mounting clip that eliminates the need for an applied glue. The weatherstripping is clipped to the inside channel of the rolled metal section of the fixed frame. To insure against galvanic corrosion between the weatherstripping (often bronze or brass), and the steel window, the window must be painted prior to the installation of the weatherstripping. Where the spring-metal type cannot be applied, use vinyl weatherstripping. This weatherstripping is usually applied to the entire perimeter of the window opening, but in some cases, such as casement windows, it may be best to avoid weatherstripping the hinge side.

2.2.2.1—Vinyl Windows

Inspect and clean vinyl windows every 2 years.

Clean vinyl windows only with mild soapy water and/or Formula 409, Ajax Liquid Cleaner, Murphy's Oil Soap, Lysol Cleaner, Soft Scrub, or vinegar and water.

Do not use any of the following products, which may cause vinyl to yellow:

- Clorox
- Pine Power
- Ivory Liquid
- Grease Relief
- Tide

- Nail Polish Remover
- Gasoline or Turpentine

Where the air space in double glazing has a "cloudy" or "milky" film, the glazing has failed and must be replaced.

2.2.2.2—Aluminum Windows and Patio Doors
Inspect aluminum windows and patio doors every 3 years.

Clean and lubricate horizontal sliders and single-hung windows as follows:
1. Remove horizontal sliding window operating sash before cleaning. It is not necessary to remove single-hung window sash for lubrication; simply open the single-hung sash all the way and apply silicone spray to the jambs as described below.

2. Vacuum track or side jambs thoroughly (check for obstructions such as rocks, pet hair, etc.).

3. Wipe with sponge, mild soap and water.

4. Rinse and let dry.

5. Apply silicone spray to dry, soft cloth; wipe on track or side jambs; do not allow silicone to come in contact with weather stripping.

6. Install sash (if previously removed).

7. Slide sash back and forth (up and down on single-hung) to check operation.

Clean and lubricate casement/awning windows as follows:
(The hardware for this window includes a handle and hinge for opening and closing the sash. The handle and hinge require regular cleaning. Periodically lubricate hinge for smooth operation. If the window seems too loose or too tight, the hinge may need adjustment.)

1. To clean and lubricate hinge, open sash and clean with a solution of mild dish soap and water., applied with a soft cloth or sponge. For stubborn dirt, use a soft bristle brush to gently scrub.

2. Rinse thoroughly, wipe dry with soft cloth, and allow to air dry.

3.   Lubricate pivot points with light oil.

4.   Adjust hinge using a small flat head screwdriver.

Clean and lubricate patio door rollers:
1.   Vacuum track thoroughly (check for obstructions such as rocks, pet hair, etc.).

2.   Clean with sponge, mild soap and water, rinse, and let dry.

3.   For light lubrication (for patio doors with rollers that rest on a track in the sill) without removing the door, apply silicone spray to cloth and wipe onto clean track

4.   For thorough lubrication, remove patio door panel, tip door panel for access to rollers on the bottom, and lubricate rollers on shaft through wheels with light oil.

Maintenance of aluminum surfaces:
1.   Inspect aluminum surfaces for scratches or cracks in the finish. Pay close attention to bare aluminum at edges and *weepholes* and areas with no finish. (Bare aluminum will oxidize rapidly in a coastal environment.) Oxidation is a natural occurrence that produces a coating that wipes off as a dark, metallic-looking residue.

2.   For optimum protection against oxidation, clean with sponge an mild soapy water and apply a coat of quality automobile wax over enamel or anodized finish on aluminum surfaces.

3.   To remove oxidation, gently scrub with fine scratch pad or steel wool; do not scratch finished surfaces, then dust or vacuum residue and wipe clean with damp cloth.

Maintenance of weep system:
1.   Clean sill track with vacuum or wipe thoroughly with damp cloth.

2.   Pour small amount of water (1 cup) into interior sill track. If water drains out through the exterior weep holes, the system is clear. If not, continue with the following steps to clean.

3.   Insert thin wire into weep hole. (*CAUTION: Some sashes have weep holes underneath the bottom rail. Do not insert wire into sash weep holes. This may damage the insulating glass seal and cause seal failure.*)

4.   Repeat until water runs clear through exterior weep hole.

Glass maintenance:
1.   Inspect glass for cracks and replace if needed.

2.   For insulating glass units, moisture between glass panes is an indication of seal failure. Replace glass.

Weather stripping maintenance:
     If weather stripping appears to be in good condition, clean as follows:
1.   Clean with damp cloth.
2.   If necessary, clean with mild soap and water.
3.   Rinse and dry thoroughly.

     Replace weather strip that is torn, cracked, brittle, discolored, gummy, or that has no "bounce back" when pressed, as follows:
1.   Carefully remove weather strip.
2.   Apply thin bead of silicone sealant into kerf.
3.   Install new weather strip.
4.   Hold in place with tape for 24 hours.
5.   Carefully remove tape.

2.2.2.3—Wood Windows
     Inspect and repair wood windows annually.

Paint maintenance:
1.   Remove paint from wood windows. Wood windows tend to accumulate many layers of paint. Over time, partial peeling leaves a pitted surface that encourages moisture to collect. The extent of paint removal required depends on the condition of the paint:
         Chalked paint: Clean with a mild detergent solvent, hose down, and allow to dry before repainting.

Crazed paint: Sand by hand to the next sound layer before repainting; exposure of bare wood is not necessary.

Peeling and blistering: Analyze between coats as to the source. If salts or impurities have caused peeling, scrape off the defective surface, hose off the underlying surface, and wipe surface dry before repainting. If the peeling or blistering was caused by incompatibility of the paints or improper application, scrape off the defective surface, and sand the underlying surface to provide a better bond with the new paint. Peeling, cracking, and alligatoring to bare wood require total removal of the defective paint followed by drying out of the wood substrate and treatment for any rotted areas before repainting. Sand or scrape only to the next sound layer of paint, exposure of bare wood is not necessary.

Remove paint from wood surfaces by scraping after it has been softened with heat guns or plates or brushed with commercially available chemical stripping solvents. Regardless which method is chosen for paint removal, after the stripping process is complete, all affected areas will need at least light sanding.

2. Once the paint has been removed, revitalize the bare wood by rubbing it with fine-grade steel wool soaked in turpentine or mineral spirits and boiled linseed oil.

3. Paint windows. Prime any exposed wood. Exterior water-based vinyl acrylic paints are generally more compatible with existing paint layers that may contain lead and provide better moisture permeability than water-or solvent-based alkyd paints. If the paint layer is impermeable, it may trap water that penetrates past the paint film. (Alkyd paints are incompatible with existing paint layers containing lead.)

Replacing glass and/or glazing compounds:
1. Broken, cracked, or missing glass must be replaced. Glazing compound that has hardened, split, or is missing must be replaced.

2. Double glazing where the air space has a "cloudy" or "milky" film has failed and must be replaced.

3.   Replace glazing compounds with an oil-based putty. Remove all deteriorated material manually by scraping, taking care not to damage the rabbet, where the glass is positioned. During all operations take every precaution to protect the glass. If the putty or other compound has hardened in the rabbets, it can be softened by applying heat. A heat gun may be used if the glass is protected by a heat shield (hardboard wrapped with aluminum foil). Before the glass panes are replaced, the surfaces of sash members should be prepared. Clean and finish bare surfaces of wood sash by rubbing the surface with a fine-grade steel wool or a fine grade of high-quality sandpaper, and then apply a solution of equal parts of boiled linseed oil and turpentine. Finally, prime and repaint.

Repairing damaged or deteriorated windows:

1.   Deteriorated portions of wood windows can be effectively repaired using like-kind splices or "Dutchmen." All deteriorated material should be removed, and the end where the replacement member will be attached should be cut on a diagonal to increase the gluing surface area. The replacement member should match the existing members in grain orientation and in any existing shape or profile. To attach the new member, cut the end diagonally to match the existing member, then drill aligned holes in both members for reinforcing dowels.

2.   A "Dutchman" is a fitted patch in a wood member that has only localized deterioration. To fit a dutchman, first probe the deteriorated area to determine the approximate depth of the deterioration. Second, cut a wood patch or dutchman with its grain aligned with the existing member's. Be sure the dutchman is large enough to cover the affected area and thicker than the deterioration is deep. Slightly bevel all of the edges of the dutchman so that the widest face is the top. Next, trace the outline of the dutchman's narrowest face on the existing member over the deterioration. Using the outline as a guide, carefully remove all of the deteriorated wood with a chisel. Test-fit the dutchman, and trim the hole until the dutchman bottoms out and fills the affected area entirely; the dutchman should be slightly higher than the existing material. Glue and clamp the dutchman in place. Once the glue has cured, use a chisel or hand plane to make the dutchman flush with the surrounding material. Hand sand for final leveling of the two surfaces.

3. When only wood surfaces are eroded, voids can be eliminated by applying a paste or putty filler. Apply fillers after the wood has dried and has been treated with a fungicide and a solution of boiled linseed oil.

4. In cases where a limited amount of rot has progressed well into the substrate, interior voids are filled in by saturating the wood with a penetrating epoxy compound formulated for wood. Fill surface voids or built up decayed or missing ends near joints with an epoxy compound.

Clean and adjust window hardware:
1. For routine cleaning use fine steel wool or a fine brass-wire brush and a cleaning solvent.

2. Moving parts should be lubricated with a non-corrosive lubricant.

2.2.2.4—Curtain Walls and Storefronts
Inspect curtain walls and storefronts annually.

The key elements of a curtain wall are the vertical steel or aluminum mullions that hold it together and the horizontal rails or mullions that are supported by the vertical members. Together, they make up the framing that support the infill panels, either vision lights (glass or plastic) and spandrel panels (solid panels of almost any material, including opaque glass).

WHAT TO LOOK FOR	REMEDIAL ACTIONS
Buckling or bending of mullions	This is usually beyond the scope of routine maintenance and may require the services of an experienced curtain wall contractor for repair.
Vertical mullions out of plumb	This is usually beyond the scope of routine maintenance and may require the services of an experienced curtain wall contractor for repair.
Loose or damaged mullions	Re-align mullions if undamaged and install new fasteners. If mullion is too damaged for continued use, replace damaged section(s) with new. If necessary, utilized the services of an experienced curtain wall contractor.

Lost or failed lights caulking at vision or spandrel panels	Remove broken, dried, or separated glazing compound. If necessary, replace glazing beads, and other fasteners and gaskets that hold the glass.
	Clean mullions thoroughly and install glass or spandrel panels with glazing compound. For steel mullions, use glazing compound specifically formulated for steel.
Corrosion of steel mullions and/or retaining bars	A sharp probe or tool, such as an ice pick, can be used to determine the extent of corrosion in the metal:
	1. If the rusting is merely a surface accumulation or flaking, then the corrosion is *light*.
	2. If the rusting has penetrated the metal (indicated by a bubbling texture), but has not caused any structural damage, then the corrosion is *medium*.
	3. If the rust has penetrated deep into the metal, the corrosion is *heavy*. Heavy corrosion generally results in some form of structural damage, through delamination, to the metal section, making repair difficult and expensive and the mullion should be replaced.
	In any case, since moisture is the primary cause of corrosion, it is essential that excess moisture be eliminated before any repair work is undertaken.
	Remove light and medium rust via manual abrasion (wire brush and aluminum oxide sandpaper) and mechanical abrasion (such as an electric drill with a wire brush or a rotary whip attachment). Rust can also be removed by using a number of commercially prepared anticorrosive acid compounds. If chemicals are used, any chemical residue should be wiped off with damp cloths, then dried immediately with Industrial blow-dryers. *Do not use running water to remove chemical residue.*
	Removing rust will remove most flaking paint as well. Remaining loose or flaking paint can be removed with a chemical paint remover or with a pneumatic needle scaler or gun. (Well-bonded paint may serve to

protect the metal further from corrosion and need not be removed. The paint edges should be feathered by sanding to give a good surface for repainting.)

Once metal has been cleaned of all corrosion, small holes and uneven areas resulting from rusting should be filled with a patching material of steel fibers and an epoxy binder and sanded smooth to eliminate pockets where water can accumulate.

Bare metal should then be wiped with a cleaning solvent such as denatured alcohol, and dried immediately in preparation for the application of an anticorrosive primer (oil-alkyd based paint rich in zinc or zinc chromate), applied immediately after cleaning.

Paint failure on steel mullions	Prepare surface and repaint in accordance with Maintenance Procedure 2.2.1.11.
Deposits or stains on aluminum mullions, surface oxidation	Inspect aluminum surfaces for scratches or cracks in the finish. Pay close attention to bare aluminum at edges and *weepholes* and areas with no finish. (Bare aluminum will oxidize rapidly in a coastal environment.) Oxidation is a natural occurrence that produces a coating that wipes off as a dark, metallic-looking residue.
	For optimum protection against oxidation, clean with sponge an mild soapy water and apply a coat of quality automobile wax over enamel or anodized finish on aluminum surfaces.
	To remove oxidation, gently scrub with fine scratch pad or steel wool; do not scratch finished surfaces, then dust or vacuum residue and wipe clean with damp cloth.
Loose, misaligned, or missing mullion covers (if used)	These trim pieces often work looses, become bent or bowed, or are knocked out of alignment. These defects will allow water to enter and require correction. Straighten and re-install existing trim, if possible. Otherwise, replace trim with new.
Rusted, loose, or missing fasteners	Steel mullion systems will use plated or coated steel fasteners that will corrode over time and must be replaced. Replace with new stainless steel fasteners.

	Aluminum mullion systems use stainless steel fasteners that may work loose over time. Replace with next larger size of stainless steel fastener.
Clogged weep holes	Weeps may become clogged with dirt, leaf litter, etc. Sometimes workmen will caulk weeps, thinking they are the source of water leaks. Pour small amount of water (1 cup) onto the base mullion. If water drains out through the exterior weep holes, the system is clear. If not, continue with the following steps to clean. Insert thin wire into each weep hole and clear it. Repeat until water runs clear through each exterior weep hole.
Damaged or color faded spandrel panels	Replace spandrel panels. Use new gaskets and/or caulking to match existing.
Broken, cracked, or missing vision lights	Replace vision lights. Use new gaskets and/or caulking to match existing. For insulating glass units, moisture between glass panes is an indication of seal failure. Replace glass.
Yellowing of plastic vision lights	Replace vision lights. Use new gaskets and/or caulking to match existing.
Damaged gaskets on vision lights or spandrel	Replace gaskets and caulking.

2.2.2.5—Interior Window Treatments

Inspect interior window blinds, shades, and shutters every 2 years. Replace failed unit(s) with new.

2.2.2.6—Exterior Doors

Inspect exterior doors every 6 months.

Frame:

Frames consist of two jamb sections, a head section, and, for double doors, an astragal in the center of the frame.

WHAT TO LOOK FOR	REMEDIAL ACTIONS
Frame anchored to wall	Look for signs of movement in the frame installation. Open the door about half way and attempt to rack the door. If frame moves under this test, or there is evidence of movement such as split caulking joints, wear, etc., the frame must be re-anchored. Wood and aluminum frames can be removed and reinstalled with new anchors. Hollow metal frames require new anchors and grouting to the wall.
Corrosion of hollow metal	A sharp probe or tool, such as an ice pick, can be used to determine the extent of corrosion in the metal: 1. If the rusting is merely a surface accumulation or flaking, then the corrosion is *light*. 2. If the rusting has penetrated the metal (indicated by a bubbling texture), but has not caused any structural damage, then the corrosion is *medium*. 3. If the rust has penetrated deep into the metal, the corrosion is *heavy*. Heavy corrosion generally results in some form of structural damage, through delamination, to the metal section, making repair difficult and expensive and the frame should be replaced. In any case, since moisture is the primary cause of corrosion, it is essential that excess moisture be eliminated before any repair work is undertaken. Remove light and medium rust via manual abrasion (wire brush and aluminum oxide sandpaper) and mechanical abrasion (such as an electric drill with a wire brush or a rotary whip attachment). Rust can also be removed by using a number of commercially prepared anticorrosive acid compounds. If chemicals are used, any chemical residue should be wiped off with damp cloths, then dried immediately with Industrial blow-dryers. *Do not use running water to remove chemical residue.*

Removing rust will remove most flaking paint as well. Remaining loose or flaking paint can be removed with a chemical paint remover or with a pneumatic needle scaler or gun. (Well-bonded paint may serve to protect the metal further from corrosion and need not be removed. The paint edges should be feathered by sanding to give a good surface for repainting.)

Once metal has been cleaned of all corrosion, small holes and uneven areas resulting from rusting should be filled with a patching material of steel fibers and an epoxy binder and sanded smooth to eliminate pockets where water can accumulate.

Bare metal should then be wiped with a cleaning solvent such as denatured alcohol, and dried immediately in preparation for the application of an anticorrosive primer (oil-alkyd based paint rich in zinc or zinc chromate), applied immediately after cleaning.

Failed paint on hollow metal or wood	Prepare surface and repaint in accordance with Maintenance Procedure 2.2.1.11.
Staining, oxidation, or pitting of aluminum	Inspect aluminum surfaces for scratches or cracks in the finish. Oxidation is a natural occurrence that produces a coating that wipes off as a dark, metallic-looking residue.

For optimum protection against oxidation, clean with sponge an mild soapy water and apply a coat of quality automobile wax over enamel or anodized finish on aluminum surfaces.

To remove oxidation, gently scrub with fine scratch pad or steel wool; do not scratch finished surfaces, then dust or vacuum residue and wipe clean with damp cloth. |
| *Bent or damaged frame* | Hollow metal or aluminum frame sections that are bent generally cannot be straightened and must be replaced and re-anchored to the wall (see above). Damaged wood frame sections must be replaced. |
| *Threshold damaged, loose, or missing* | Re-anchor loose threshold.
Replace damaged or missing threshold. |

Removable astragal damaged or loose	Fixed astragals are part of the door frame. Removable astragals must be straight and tight-fitting to form a weatherproof element of the door opening. If damaged, replace the astragal. If loose, repair as necessary for tight door fit.

Door:

WHAT TO LOOK FOR	REMEDIAL ACTIONS
Door racked or warped, bent	Check that door fits flush against frame stops on all three sides. If not, door fit may be adjusted by adjusting or moving hinges. For severe racking or warp, door replacement is required.
Loose fit	Loose fit is caused by missing or failed weather stripping, bent or damaged frame, loose hinges, etc. Check each item and repair as necessary.
Rust on metal	If corrosion is extensive, especially at the bottom of the door, it must be replaced. Light rusting can be repaired as follows.
	Remove rust via manual abrasion (wire brush and aluminum oxide sandpaper) and mechanical abrasion (such as an electric drill with a wire brush or a rotary whip attachment). Rust can also be removed by using a number of commercially prepared anticorrosive acid compounds. If chemicals are used, any chemical residue should be wiped off with damp cloths, then dried immediately with Industrial blow-dryers. *Do not use running water to remove chemical residue.*
	Removing rust will remove most flaking paint as well. Remaining loose or flaking paint can be removed with a chemical paint remover or with a pneumatic needle scaler or gun. (Well-bonded paint may serve to protect the metal further from corrosion and need not be removed. The paint edges should be feathered by sanding to give a good surface for repainting.)

	Once metal has been cleaned of all corrosion, small holes and uneven areas resulting from rusting should be filled with a patching material of steel fibers and an epoxy binder and sanded smooth to eliminate pockets where water can accumulate.
	Bare metal should then be wiped with a cleaning solvent such as denatured alcohol, and dried immediately in preparation for the application of an anticorrosive primer (oil-alkyd based paint rich in zinc or zinc chromate), applied immediately after cleaning.
Wet or dry rot of wood	Test wood condition of door, especially near the bottom, with a screwdriver or ice pick to determine if door has wet or dry rot. Small areas of rot can be removed and the door patched with a Dutchman. Larger areas can be treated with epoxy coatings and filler. After repairs, prepare surface and repaint in accordance with Maintenance Procedure 2.2.1.11.
	Extensive rot cannot be repaired and the door must be replaced.
	Wet rot usually results from poor door weather protection and/or failed paint. Check doorway surround for rain or flooding problems that has damaged door. Correct these problems!
Failed paint on metal or wood	Prepare surface and repaint in accordance with Maintenance Procedure 2.2.1.11.
Staining, oxidation, or pitting of aluminum	Inspect aluminum surfaces for scratches or cracks in the finish. Pay close attention to bare aluminum at edges and *weepholes* and areas with no finish. (Bare aluminum will oxidize rapidly in a coastal environment.) Oxidation is a natural occurrence that produces a coating that wipes off as a dark, metallic-looking residue.
	For optimum protection against oxidation, clean with sponge an mild soapy water and apply a coat of quality automobile wax over enamel or anodized finish on aluminum surfaces.

	To remove oxidation, gently scrub with fine scratch pad or steel wool; do not scratch finished surfaces, then dust or vacuum residue and wipe clean with damp cloth.
Impact damage	If possible, repair damage and refinish door. If damage is severe, replace the door.
Kickplate damaged, loose, or missing	Re-anchor loose kickplate. Replace damaged or missing kickplate.

Glazing:

WHAT TO LOOK FOR	REMEDIAL ACTIONS
Broken, cracked, or missing glazing	Replace glazing. Use new gaskets and/or caulking to match existing. For insulating glass units, moisture between glass panes is an indication of seal failure. Replace glass.
Loose stops/beads	Replace.
Gasket or glazing compound defects	Remove broken, dried, or separated glazing compound. If necessary, replace glazing beads, and other fasteners and gaskets that hold the glass. Clean opening thoroughly and install glass with glazing compound. (For hollow metal doors, use glazing compound specifically formulated for steel.)

Hardware:

WHAT TO LOOK FOR	REMEDIAL ACTIONS
Hinges broken or bent	Replace hinges with new fasteners.
Latch, lock, or bolt not functional	Test operation and repair or replace as necessary. Check and tighten lock screws as necessary. Lubricate with dry graphite lubricant as needed.

*Closer/holdopen functional*closing.	Closer can be adjusted for rate of swing and force of Adjust force of closing so that door will close tightly against its jamb stop and the latch/lock engages. In high wind areas, adjust closer so door will not open too far.
Panic bar damaged, loose, or not functional	Test operation and repair or replace as necessary.
Stop damaged, loose, or missing	Re-anchor loose stop. Replace damaged or missing stop.
Weather stripping condition	Replace door weather stripping and seals every 5 years.
Automatic operator	See Maintenance Procedure 2.2.2.7
Fire/smoke hold-open devices	Test for proper function. Repair or replace as necessary.

2.2.2.7—Exterior Door Automatic Operators

Inspect and repair exterior door automatic operators in accordance with Appendix H.

2.2.2.8—Overhead Doors

Inspect overhead doors every 6 months.

WHAT TO LOOK FOR	REMEDIAL ACTIONS
Track anchors, alignment, and condition	Check anchors for looseness, missing or corroded fasteners, etc. and repair as necessary.
	Check track alignment. Adjust anchors as need. If track has been damaged, replace track section.
	Check for dirt or other material (metal shavings, sawdust, etc.) that can clog the track and clean as needed.

	Check for corrosion. Remove rust via manual abrasion (wire brush and aluminum oxide sandpaper) and mechanical abrasion (such as an electric drill with a wire brush or a rotary whip attachment). Rust can also be removed by using a number of commercially prepared anticorrosive acid compounds. If chemicals are used, any chemical residue should be wiped off with damp cloths, then dried immediately with Industrial blow-dryers. *Do not use running water to remove chemical residue.*
	Bare metal should then be wiped with a cleaning solvent such as denatured alcohol, and dried immediately in preparation for the application of an anticorrosive primer (oil-alkyd based paint rich in zinc or zinc chromate), applied immediately after cleaning. Paint in accordance with Maintenance Procedure 2.2.1.11.
Rollers	Rollers must run in the track, requiring that roller pins be properly aligned and anchored. Adjust or replace as necessary. Rollers that do not turn easily or show significant corrosion or damage must be replaced.
Door damaged or racked	Repair or replace damaged door panels as necessary. Check that door fits flush on all three sides and tight against the floor. If not, door fit may be adjusted by adjusting or moving rollers or adjusting track alignment. For severe racking or warp, door replacement is required.
Paint failure	Prepare surface and paint in accordance with Maintenance Procedure 2.2.1.11. If metal door panels have light rust, treat as for tracks, above.
Lock/latch	Test operation and repair or replace as necessary.
Electric operator	Test for proper operation in accordance with manufacturer's maintenance and operation instructions. Repair as necessary.

2.2.2.9—Cafeteria and Kitchen Air Curtains

Inspect cafeteria and kitchen air curtains every 3 months. Test for proper operation and repair or replace as indicated by test.

2.3.1.0—Low-slope Roof Housekeeping

General "housekeeping inspection" of a low-slope roof is required every month to address the following elements:

1.   Remove trash, leaves, tree limbs, sports balls, drink cans and bottles, workmen's tools, etc. from the roof.

2.   Check/clean gutters, downspouts, roof drains, and roof scuppers. Repair/replace damaged/missing roof drain dome strainers.

3.   Check for open roof hatches, open doors or panels on rooftop HVAC equipment, exhaust fans with broken belts (evidenced by motor running but no airflow), exhaust fans and gravity hoods, etc. with damaged housings, broken vent pipes, etc.

4.   Check for wind or storm damage to roof covering system elements such as flashing, parapet caps, cant strips, standing seam panels, etc.

5.   If roof is ballasted, check condition of ballast, particularly at corners and along edges where wind damage is most likely to occur.

6.   Record any ponding that exists and add this record to the semi-annual condition inspection report.

7.   If tree limbs overhang the roof, trim them back to where they will remain clear of the roof under high wind conditions.

2.3.1.1—Roofs

All roofs must be inspected after a storm, wind, and/or hail event for damage; leaves, limbs, and other debris on the roof; missing or damaged roofing elements; and water leaks.

Make temporary repairs to stop water intrusion immediately. Initiate process for making permanent repairs or replacement. (Hail damage repairs may be covered under the schools insurance program.

Advise Facilities Director immediately if hail damage is discovered.)

For sloped roofs (pitch greater than 3/12), the inspection maintenance requirements shall comply with Maintenance Procedures 2.3.1.2.1 through 2.3.1.2.6, as applicable.

For low-slope roofs (pitch less than 3/12), the inspection and maintenance requirements shall comply with Maintenance Procedures 2.3.1.3.1 and 2.3.1.3.2.

2.3.1.2.1—Sloped Asphalt Shingle Roofs

Sloped asphalt shingle roofs must be inspected every 6 months to evaluate roof condition, define immediate maintenance needs, and adjust replacement life time estimates.

WHAT TO LOOK FOR	PROBABLE CAUSE	REMEDIAL ACTIONS
Shingle surface wear, including loss of aggregate granules, exposure of underlying fiberglass mesh, ragged edges, deformation or "sagging" or "clawing," etc.	Deterioration due to age, heat, water and wind, etc. Hail damage.	With minor damage, where only limited areas from which the coating and surfacing granules have been loosened, may be repaired by covering the bare areas with asphalt-base roof coating, plastic cement, or clay type asphalt emulsion. Where the asphalt coating and surfacing granules have been lost from numerous small areas, but the shingles are not broken, will not cause the roof to leak. However, the life of a roof so damaged will be shortened materially...adjust remaining life estimate accordingly. Severe hail storms may damage asphalt-shingle

		roofs beyond repair, particularly if the shingles have been exposed for a number of years. With such damage, both layers of shingles are broken and the roof will leak severely. Re-roofing is mandatory. During each inspection, evaluate shingle condition and adjust anticipated remaining performance life accordingly.
Shingle holes, breaks, or splits	Mechanical damage due to tree limbs, hail, or roof traffic.	Holes, small breaks, and minor hale damage can be repaired by applying asphalt plastic cement. Larger damaged areas require replacement of shingles.
Lifted tabs	Typically due to adhesive failure or improper nailing. Wind damage.	Place a spot of quick setting asphalt plastic cement under the center of each tab (two spots for each tab of 2-tab shingles) and pressing the tab down firmly. The spot of cement pressed flat. Approximately ½ gallon of cement is required per square of shingles. The shingle tabs should not be bent up farther than necessary to place the cement. *No attempt should be made to re-nail shingles.*
Water damage to sheathing, facia, and/or soffit	Lack of metal drip edge and proper shingle overhang.	Where no metal drip edge has been installed, check for rotted wood at the roof eaves (sheathing, facia, and soffits).

		If no rotted wood is found, install new metal drip edge. Carefully remove nails holding the shingles at the rake and then install a new galvanized metal drip edge. The shingles should then be re-nailed and all of the shingle tabs adjoining the rake cemented as described above.
		If rotted wood is found, the first three or four courses of shingles must be removed and deteriorated sheathing, facia, and/or soffits replaced. In replacing the shingles, a new galvanized metal drip edge is required. Particular care should be taken not to damage the old shingles when removing nails to join the new roof section with the old one.
Moisture infiltration	Flashing problems at chimneys, skylights, HVAC equipment, etc. including rust, separation, lifting, etc.	Replace deteriorated or damaged flashing materials. Replace rubber boots on plumbing vents as required.
	Roof penetrations are the primary source of roof leaks. Check carefully for open seams, failed rubber boots, failed metal flashing, etc.	Split or leaking valley flashing can be replaced with metal, asphalt roll roofing, or by close woven shingles. In most cases, metal is preferred.
	Check all roof valleys and ridge vents.	Do not use galvanized steel for replacement of metal flashing or counter flashing...use copper or stainless steel, with matching fasteners.

2.3.1.2.2—Sloped Asphalt Roll Roofing

Sloped asphalt roll roofing must be inspected every 6 months to evaluate roof condition, define immediate maintenance needs, and adjust replacement life time estimates.

WHAT TO LOOK FOR	PROBABLE CAUSE	REMEDIAL ACTIONS
Roofing surface wear, including loss of aggregate granules and exposure of underlying fiberglass mesh.	Deterioration due to age, heat, water and wind, etc. Hail damage.	If surface wear is significant, the roof should be replaced. Otherwise, no action is required. During each inspection, evaluate roof condition and adjust anticipated remaining performance life accordingly.
Splits, holes, or breaks	Mechanical damage due to tree limbs, hail, or roof traffic.	Holes, small breaks, and minor hale damage can be repaired by applying asphalt plastic cement. Larger areas can be repaired by opening the horizontal seam below the break and inserting through it a strip of roofing material. Extend the strip at least 6 inches beyond the edges of the break, with the lower edge flush with the horizontal exposed edges of the covering sheet. Coat the strip liberally with lap cement where it will come into contact with the covering sheet before inserting it. After inserting the strip, press down the edges of the roofing firmly and nail

		with about 3/4" from the edges, 2" on center. Apply lap cement to the horizontal seam, press down firmly, and re-nail.
		If more than 25% of the roof area requires repair, replace the entire roof.
Leaky seams	Insufficient lap, inadequate nailing or nail pull through, wind damage, or roof traffic.	Use asphalt saturated woven glass fabric cemented over the seam and coated with bituminous compound. Apply the coating to the seams in strips approximately 6" wide, using about 1 gallon of coating for each 80 linear feet of seam. Embed a 4" wide strip of saturated fabric in the coating, centered over the seam, pressing it firmly into the coating until it lies flat without wrinkles or buckles. Then, apply another layer of coating and a second strip of saturated fabric so that both coatings are continuous.
Water damage to sheathing, facia, and/or soffit	Lack of metal drip edge and proper roofing overhang.	Where no metal drip edge has been installed, check for rotted wood at the roof eaves (sheathing, fascia, and soffits).
		If no rotted wood is found, install new metal drip edge. Carefully remove nails holding the roofing at the rake and then install a new

		galvanized metal drip edge. Re-nail roofing with 3/4" nails, 4" on center.
		If rotted wood is found, remove roll roofing 2-3 feet beyond the deteriorated sheathing, facia, and/or soffits. Replace wood and install new galvanized metal drip edge. Install new roll roofing with at least 4" laps over old roofing and nail with 3/4" nails, 4" on center.
Moisture infiltration	Flashing problems at chimneys, skylights, HVAC equipment, etc. including rust, separation, lifting, etc.	Replace deteriorated or damaged flashing materials. Replace rubber boots on plumbing vents as required.
	Roof penetrations are the primary source of roof leaks. Check carefully for open seams, failed rubber boots, failed metal flashing, etc.	Split or leaking valley flashing can be replaced with metal or asphalt roll roofing.
	Check all roof valleys and ridge vents.	Do not use galvanized steel for replacement of metal flashing or counter flashing...use copper or stainless steel, with matching fasteners.

2.3.1.2.3—Sloped Wood Shake or Shingle Roofs

Sloped wood shake or shingle roofs must be inspected every 3 months to evaluate roof condition, define and perform immediate maintenance needs, determine permanent roof modifications that may be required, and adjust replacement life time estimate.

WHAT TO LOOK FOR	PROBABLE CAUSE	REMEDIAL ACTIONS
Weathering	Sunlight and moisture will turn the shingles gray that darkens over time.	This is a natural condition and requires no action. Wood preservative can be applied to new roofs within 3-12 months of installation and then annually thereafter. (Treatment of older roofs that have never been treated before will not be effective.)
Mold, mildew, algae, and or moss growth	Organics and moisture is trapped between shingles and in roof valleys	Clean roof 3-4 times per year to remove leaf litter, pine nettles, and other organic material. Wash off the material with a garden hose, using a stiff broom as needed. A cleaning solution of 3 oz. TSP, 1 oz detergent, and 1 qt of laundry bleach mixed with 3 quarts of warm water should be used. Pressure washing is an option. But, care must be taken keep the pressure low so as not to mechanically damage shingles.
Leaks	Holes, large splits, water barrier failure, etc.	Replacement of the roof in the leak area is required. Check for wood rot in the sheathing and replace as needed.
Water damage to sheathing, facia, and/or soffit	Lack of metal drip edge and proper shingle overhang.	Where no metal drip edge has been installed, check for rotted wood at

		the roof eaves (sheathing, facia, and soffits).
		If no rotted wood is found, install new metal drip edge. Carefully remove nails holding the shingles at the rake and then install a new galvanized metal drip edge. The shingles should then be re-nailed.
		If rotted wood is found, the first three or four courses of shingles must be removed and deteriorated sheathing, facia, and/or soffits replaced. In replacing the shingles, a new galvanized metal drip edge is required. Particular care should be taken not to damage the old shingles when removing nails to join the new roof section with the old one.
Moisture infiltration	Flashing problems at chimneys, skylights, HVAC equipment, etc. including rust, separation, lifting, etc.	Replace deteriorated or damaged flashing materials. Replace rubber boots on plumbing vents as required.
	Roof penetrations are the primary source of roof leaks. Check carefully for open seams, failed rubber boots, failed metal flashing, etc.	Split or leaking valley flashing can be replaced with metal, asphalt roll roofing, or by close woven shingles. In most cases, metal is preferred.
	Check all roof valleys and ridge vents.	Do not use galvanized steel for replacement of metal flashing or counter flashing...use copper or stainless steel, with matching fasteners.

2.3.1.2.4—Sloped Slate Tile Roofs

Sloped slate tile roofs must be inspected annually to evaluate roof condition, define and perform immediate maintenance needs, determine permanent roof modifications that may be required, and adjust replacement life time estimate.

WHAT TO LOOK FOR	PROBABLE CAUSE	REMEDIAL ACTIONS
Damaged, missing or loose slate tiles	Mechanical damage due to hail, roof traffic, falling limbs, warping of the roof, etc. Fastener failure	If only a few slates are broken, missing, or loose, they should be removed and new ones applied. If 20-25% or more are broken, missing, or loose, all slates should be removed, salvaged, and new roof installed, using salvaged slates plus matching new ones mixed in, since this will be cheaper than individual slate repairs. To replace a small number of slates, remove the broken or loose slate and cut the nails with a ripper. Insert a new slate and nail through the vertical joint of the next course above, driving the nail about 2" below the butt of the slate in the second course above. For water protection, force a 3" x 8" strip of copper under the course above the nail and bend the strip slightly concave to hold it in place. The strip should extend about 2" under the second course and cover the nail,

		extending 2" below it. Use copper or stainless steel nails, only.
Water damage to sheathing, facia, and/or soffit	Lack of metal drip edge and proper shingle overhang.	Where no metal drip edge has been installed, check for rotted wood at the roof eaves (sheathing, facia, and soffits). If no rotted wood is found, install new metal drip edge. Carefully remove nails holding the slates at the rake and then install a new galvanized metal drip edge. The slates are then reinstalled as described above. If rotted wood is found, the first three or four courses of shingles must be removed and deteriorated sheathing, facia, and/or soffits replaced. In replacing the slates, a new galvanized metal drip edge is required. Particular care should be taken not to damage the old slates when removing nails to join the new roof section with the old one.
Moisture infiltration	Flashing problems at chimneys, skylights, HVAC equipment, etc. including rust, separation, lifting, etc. Roof penetrations are the primary source of roof leaks. Check carefully for	Replace deteriorated or damaged flashing materials. Replace rubber boots on plumbing vents as required. Split or leaking valley flashing can be replaced with metal or asphalt roll

open seams, failed rubber boots, corroded metal flashing, etc. Check all roof valleys and ridge vents.	roofing. In most cases, metal is preferred. Do not use galvanized steel for replacement of metal flashing or counter flashing...use copper or stainless steel, with matching fasteners.

2.3.1.2.5—Sloped Metal Roofs

Sloped (pitch greater than 3/12) metal roofs must be inspected every 6 months to evaluate roof condition, define immediate maintenance needs, and adjust replacement life time estimates.

Typically, three types of metal roofs are utilized:

- *Corrugated or "Five-V" galvanized* steel applied with external fasteners.

- *Structural metal* with baked enamel finish used on pre-engineered metal buildings that typically employ 24 to 26 gauge coated steel panels installed over structural support. Steel panels may be corrugated for additional rigidity and aesthetic reasons. The steel panels are fastened to the structural support at their sidelap and/ or endlap joints. One method used to connect ribbed panels is with self-drilling screws, nuts and bolts or rivets. With this type of system, gasket washers are used to provide weatherproofing. Panel joints are sealed and waterproofed during installation using butyl rubber sealants and tape.

- *Architectural standing seam roofing,* with a baked enamel finish, applied over roof structure, deck, and insulation. The metal most commonly used is "galvalume," aluminum-coated galvanized steel. Panels are joined together by a weathertight seam that is raised above the roof's drainage plane. Factory-applied organic sealants are applied to the seam during roll forming of the panel. Once on site, automatic field seaming machines complete the seal. Metal panels are attached to the roof substructure with concealed clips. These are roll formed or crimped into the panel seams without penetrating the corrosion-resistant steel weathering membrane. The clip system performs two functions: it retains the panels in position without exposed fasteners and allows the roof to expand and contract during temperature changes.

Galvanized steel corrugated or 5-V roofing:

WHAT TO LOOK FOR	PROBABLE CAUSE	REMEDIAL ACTIONS
Corrosion	General rusting indicates that the zinc coating is no longer protecting the steel. Localized rusting, particularly at fasteners, indicates galvanic corrosion.	To protect the steel, it should be painted every 3-5 years. New surfaces require a coats of wash primer to remove the mill finish, zinc oxide primer, and finish paint. Repainting requires removal of rust by sanding, priming with red lead or other rusty metal primer, and application of a finish coat. Localized rusting requires fastener replacement with rubber grommets and washers to eliminate contact between roof and fastener. Remove rust by sanding, priming with red lead or other rusty metal primer, and application of a finish coat before installing fasteners.
Splits, holes, or breaks	Mechanical damage due to wind, tree limbs, hail, or roof traffic.	Replace the defective sheet with a new one. Paint to match existing. In high wind areas, install additional fasteners in edge panels.
Water leaks	Leaks at seams.	Inadequate laps in galvanized steel roofing

		may be repaired by caulking opened seams or, in severe cases, stripping the seam as follows:
		Open the horizontal seam below the break and insert through it a strip of roofing material. Extend the strip at least 6 inches beyond the edges of the break, with the lower edge flush with the horizontal exposed edges of the covering sheet and cover any exposed fasteners. Coat the strip liberally with lap cement where it will come into contact with the covering sheet before inserting it. After inserting the strip, press down the edges of the roofing firmly and nail with about 3/4" from the edges, 2" on center. Apply lap cement to the horizontal seam, press down firmly, and re-nail.
	Neoprene washers are widely used at fasteners, but dry and crack over time.	Fastener leaks due to washer failure can be repaired by replacing washers with EPDM type.
Water damage to sheathing, fascia, and/or soffits	Insufficient overhang, allowing water get under the roof at the eaves	Remove existing roofing, replace rotted wood, and install roofing with at least 1-1/2" overhang on all edges.

Moisture infiltration	Flashing problems at chimneys, skylights, HVAC equipment, etc. including rust, separation, lifting, etc.	Replace deteriorated or damaged flashing materials. Replace rubber boots on plumbing vents as required.
	Roof penetrations are the primary source of roof leaks. Check carefully for open seams, failed rubber boots, failed metal flashing, etc.	Split or leaking valley flashing can be replaced with metal to match existing.
	Check all roof valleys and ridge vents.	

Structural metal roofing:

WHAT TO LOOK FOR	PROBABLE CAUSE	REMEDIAL ACTIONS
Corrosion	General rusting indicates that the paint finish and zinc coating is no longer protecting the steel.	Thoroughly clean roof, remove rust by sandblasting or sanding, apply zinc primer and final coat of acrylic elastomeric coating (5-10 mils thick).
	Scratches and dings.	Clean, sand, and touch-up areas with manufacturer provided touch-up materials.
	Localized rusting, particularly at fasteners, indicates galvanic corrosion.	Localized rusting requires fastener replacement with rubber/EPDM grommets and washers to eliminate contact between roof and fastener. Remove rust by sanding, priming with

		zinc chromate or other rusty metal primer, and application of a finish coat to match existing before installing fasteners.
	Chemical contamination.	Ferrous debris from maintenance activities, copper from HVAC condensate drains, and contact with galvanized steel counter-flashing must be corrected.
Splits, holes, or breaks	Mechanical damage due to wind, tree limbs, hail, or roof traffic.	Replace damaged roof panel(s).
	Panel lift due to wind at eaves due to inadequate fascia protection.	*Contact manufacturer about replacing fascia to provide better wind resistance.*
Water leaks (indicated by interior water stains or sagging roof insulation)	Leaks at seams.	Use a feeler gauge to determine if sealant is present under the endlaps on the wet side of fasteners. Inadequate laps or poor sealing can be repaired by caulking opened seams or, in severe cases, stripping the seam as follows:
		Open the horizontal seam below the break and insert through it a strip of roofing material. Extend the strip at least 6 inches beyond the edges of the break, with the lower edge flush with the horizontal exposed edges of the covering sheet and cover any exposed

		fasteners. Coat the strip liberally with lap cement where it will come into contact with the covering sheet before inserting it. After inserting the strip, press down the edges of the roofing firmly and nail with about 3/4" from the edges, 2" on center.
	Neoprene washers are widely used at fasteners, but dry and crack over time.	Fastener leaks due to washer failure can be repaired by replacing washers with EPDM type in lieu of original neoprene washers.
Moisture infiltration	Flashing problems at chimneys, skylights, HVAC equipment, etc. including rust, separation, lifting, etc.	Replace deteriorated or damaged flashing materials. Replace rubber boots on plumbing vents as required.
	Roof penetrations are the primary source of roof leaks. Check carefully for open seams, failed rubber boots, failed metal flashing, etc.	Split or leaking valley flashing can be replaced with metal roofing to match existing.
	Check all roof valleys and ridge vents.	

Architectural standing seam metal roofing:

WHAT TO LOOK FOR	PROBABLE CAUSE	REMEDIAL ACTIONS
Corrosion	General corrosion is caused by deterioration of finishes and exposure of base steel to air.	Clean by sandblasting or sanding, followed by rust inhibiting primer and topcoat. (For PVDF coatings, adhesion of retreatments is a

		problem.) *Consult with manufacturer for suitable primers and coatings.*
	Localized corrosion can be caused by localized finish failure (scratches, dings, etc.) and/or galvanic action.	Clean, sand, and touch-up areas with manufacturer provided touch-up materials.
	Chemical contaminants.	Ferrous debris from maintenance activities, copper from HVAC condensate drains, and contact with galvanized steel counter-flashing must be corrected.
Splits, holes, or breaks	Mechanical damage due to wind, tree limbs, hail, or roof traffic.	Temporarily patch holes or splits with embedded polyester mat or butyl tape, followed by screwing a piece of metal panel into place. Permanent fix will require panel replacement. Contact manufacturer authorized roofing contractor.
Seam leakage	Separation, failure of sealant in seam.	Temporarily patch with embedded polyester mat or butyl tape, followed by coating. Permanent fix will require that the seal be disassembled and resealed.
Panel deformation or other signs of excessive movement	Deformation, ripples ("oil canning"), etc. are generally caused by inadequate expansion allowance.	Contact manufacturer authorized roofing contractor.

| Moisture infiltration | Flashing problems at chimneys, skylights, HVAC equipment, etc. including rust, separation, lifting, etc.

Roof penetrations are the primary source of roof leaks. Check carefully for open seams, failed rubber boots, failed metal flashing, etc.

Check all roof valleys and ridge vents. | Replace deteriorated or damaged flashing materials. Ensure that flashing provides ample movement allowance. Replace rubber boots on plumbing vents as required. Do not use copper for flashing or counter-flashing.

Split or leaking valley flashing can be replaced with metal to match existing. |

2.3.1.2.6—Sloped Built-up Bitumen Membrane Roofs

Sloped built-up bitumen membrane roofs must be inspected every 6 months to evaluate roof condition, define and perform immediate maintenance needs, determine permanent roof modifications that may be required, and adjust replacement life time estimate.

These roofs consist of pre-manufactured sheet membranes called "felts" consisting of asphalt, modified with a polymer, which improves the physical properties of the asphalt and coated on a reinforcing mat or carrier. Sheets are installed in hot asphalt, typically, in 3 or 5 plies, and covered with a gravel topping.

WHAT TO LOOK FOR	PROBABLE CAUSE	REMEDIAL ACTIONS
Surface wear, including loss of aggregate granules, exposure of underlying fiberglass mesh, etc.	If roof pitch exceeds 3/12, water flow across the roof surface with erode the surface and felts. Deterioration due to age, heat, water and wind, etc. Hail damage.	If the roof pitch is greater than 3/12, roof should be replaced with longer life roof. If the roof pitch is 3/12 or less, see Maintenance Procedure 2.3.1.3.2.

2.3.1.3.1—Low-sloped Flexible or Single-ply Membrane Roofs

Low-sloped flexible or single-ply membrane roofs must be inspected every 6 months, Spring and Fall, to evaluate roof condition, define and perform immediate maintenance needs, determine permanent roof modifications that may be required, and adjust replacement life time estimate.

There are two typical types of single-ply membranes

- *EPDM (Ethylene Propylene Diene Monomer):* A generic description for synthetic rubber sheet membranes. Applied only in single-ply applications on all roof types. Seams are sealed with proprietary adhesives.

- *PVC (Polyvinyl Chloride) and TPO (Thermoplastic polyolefin):* Generic descriptions for a plastic sheet membrane. Applied in single-ply applications on all roof types. Seams are sealed by fusion, either with solvent or hot-air welding techniques.

WHAT TO LOOK FOR	PROBABLE CAUSE	REMEDIAL ACTIONS
Cuts and punctures	Inadequate or incorrectly located walk paths. Too much roof traffic. Poor work practices by HVAC, electrical, etc. maintenance staff.	Small cuts in a roofing membrane are difficult to detect if the membrane is covered with a layer of dust or dirt. If cuts and punctures are suspected, the membrane should be cleaned with water and inspected. Most cuts and punctures will produce small bubbles in a film of water applied to the surface of the membrane. Most single-ply cuts and punctures can be repaired temporarily by covering the cut or puncture with duct tape. Clean the membrane around the cut with a non-abrasive cleaner and apply a piece of duct tape extending

		beyond the cut or puncture at least one inch in all directions.
Compressed or Crushed Roof Insulation	Most roof insulations have relatively low compressive strengths and can easily be compressed or crushed if traffic loads exceed the strength of the insulation. Crushed insulation can be indicated by the presence of ponding water and "tented" insulation fasteners which did not compress along with the insulation.	Insulation that has been crushed should be replaced with new insulation. Because this procedure will also require extensive repairs to or replacement of the roofing membrane, this work must be performed by a manufacturer authorized contractor. Add walk pads along "natural" paths to protect roof membrane and insulation.
General Cracking, Crazing, Or Splitting Membrane	Typically caused by a defective membrane.	*Contact the manufacturer immediately.*
Ridging or Buckling of Membrane at Insulation Joints	Caused, typically, by movement of the substrate due to moisture or thermal expansion/contraction.	*Contact the manufacturer immediately.*
Membrane swelling, splitting, or cracking around air conditioning compressors, kitchen exhaust fans, fume hood vents, etc.	Some single-ply membranes, such as EPDM, will exhibit swelling or buckling of the membrane that typically is an indication that the membrane has absorbed an atmospheric or chemical contaminant and that fundamental physical properties of the membrane have been compromised. Contact with oils or organic solvents can embrittle some single-ply membranes by accelerating the loss of lighter weight components.	Membrane which has swelled or cracked should be replaced by a manufacturer authorized contractor. Replace degraded EPDM in these areas with a manufacturer recommended chemically resistant system such as polyepichlorohydrin or neoprene. Avoid bituminous repair materials.

General membrane discoloration	Early indication of chemical or atmospheric contamination and membrane damage.	Contact the manufacturer immediately.
Ponding (Although improper roof drainage can best be observed immediately after a rain storm, most impacted drainage conditions will leave "tell-tale" indications even after standing water has evaporated).	Because most horizontal structural members deflect in the center of the span, ponded areas are located frequently along the mid-span of these framing members. Heavy rooftop HVAC units can frequently cause deck deflection and create a ponding area around the unit. Clogged or failed drains, drainage piping, gutters and downspouts, etc.	Add additional tapered insulation in the low areas. Add additional drains if possible.
Moisture infiltration	Water entry at building walls and parapets, rooftop equipment, and skylights. Typical problem conditions are indicated by the following: 1. "Soft" Roof Insulation: If the roof insulation appears to be "soft" under foot, it may have absorbed excessive moisture. 2. Cracking, Spalling or Discoloration of Walls: The deterioration may be an indication of moisture entry. 3. Loose or Corroded Metal Wall Flashings: Look for any discontinuities in the firm, uniform compression between metal flashings and the wall surface.	Replace wet roof insulation. Re-attach and re-caulk flashing and counter-flashing components. Repair deteriorated walls or parapets.

	4. Plugged "Weep Holes": Look for masonry weep holes that have become clogged or were accidentally covered over by the roof flashings.	
	5. Missing or Broken Weather Seals on Equipment Housings: Frequent maintenance of rooftop equipment may allow weather seals and sheet metal joints to loose water-tightness.	
	6. Cracked or Sunken Caulking: Any cracking or other discontinuity is a potential source of water entry.	
Roof membrane seams separation	The leading edge of the seam may be starting to open up, allowing dirt to accumulate in a cavity at the seam edge. This typically occurs at two locations: 1. At "T" joints that occur where two sheets of roofing membrane intersect. Because of the extra thickness of membrane at these locations, these joints may over time begin to open up due to the "memory" of the membrane. 2. At angle changes. Roof seams which travel through an angle change, such as a deck-to-wall joint, are subject to the same long-term stresses as "T" joints, and may begin to open up over time.	Single-ply field seams can be repaired temporarily by covering the seam edge with duct tape. Clean the membrane around the edge with a non-abrasive cleaner and apply a piece of duct tape extending beyond the affected area at least one inch in all directions. Restoration of aged roofing seams should be performed by a manufacturer-authorized roofing contractor.

| *Membrane attachment separation* | "Bridging" refers to the tendency for roofing membrane to pull away from any angled intersection due to the inherent "memory" of the manufactured roofing sheet. Minor bridging, extending beyond the angle change less than one inch, can typically be expected from any roofing membrane after a reasonable period of service. Bridging greater than one inch may indicate that the base attachment | If bridging or distortion at a base attachment is causing active leakage, try to seal the leak with a construction-grade butyl caulking.

 Contact manufacturer authorized contractor for permanent repairs. |

2.3.1.3.2—Low-sloped Built-up Bitumen Membrane Roofs

Low-sloped built-up bitumen membrane roofs must be inspected every 6 months to evaluate roof condition, define and perform immediate maintenance needs, determine permanent roof modifications that may be required, and adjust replacement life time estimate.

These roofs consist of pre-manufactured sheet membranes called "felts" consisting of asphalt, modified with a polymer, which improves the physical properties of the asphalt and coated on a reinforcing mat or carrier. Sheets are installed in hot asphalt, typically, in 3 or 5 plies, and covered with a gravel topping.

WHAT TO LOOK FOR	PROBABLE CAUSE	REMEDIAL ACTIONS
Traffic damage	Inadequate or incorrectly located walk paths. Too much roof traffic. Poor work practices by HVAC, electrical, etc. maintenance staff.	Repair/seal cuts or punctures. Add walk pads along "natural" paths to protect roof membrane. Insulation which has been crushed should be replaced with new insulation, which will

		also require extensive repairs to the roofing membrane.
Blueberry blisters in surface of bitumen	Expansion of volatile fractions of Bitumen or of air or water, in sunny weather. More common with low meltpoint bitumens particularly with heavy coatings and poor gravel cover.	Initially apply additional opaque gravel. If many blisters occur and are broken to expose felts, recoat with bitumen and apply heavy layer of opaque gravel to prevent re-occurrence.
Blisters between layers of felt	Expansion in sunny weather of entrapped air or water in areas of poor adhesion.	Cut blister, trim excess material, re-adhere and patch. Add heavy layer of opaque gravel to prevent re-occurrence.
Blisters between felt membrane and substrate	Expansion in sunny weather of entrapped air or water, usually over decks with concrete fills, or with wetted insulation.	Where possible, cut blister, trim excess material, re-adhere and patch. Venting (if possible) may help. Add heavy layer of opaque gravel.
Ridging or buckling	Movement of either the felts or the deck or substrate under moisture or thermal effects, causing long ripple ridges especially where felt not well bonded, often over insulation or deck joints.	Usually little can be done. If of small size and elevation, a heavy application of gravel will make it less conspicuous and give some protection. If wide and high, cutting and relaying is necessary. If felt edges are exposed, scrape off all surfacing material for 2.5' beyond exposed edges, removed failed felts, and install new felts and gravel topping.
Cracking or breaking	Breaks in unsupported felt. Cracking of blisters or ridges by traffic. Breaks at sharp bends in felt.	Cut out, provide support and patch. Cut blister, re-adhere felts and patch. Round off and patch.

Felt penetrating top pour and gravel at laps	Poor workmanship initially resulting in poor adhesion. Curling of felt edges when left exposed too long during construction.	Re-adhere if possible, or cut away. Cover with flood coat of bitumen and gravel. Felt edge must not be exposed. If felt edges are exposed, scrape off all surfacing material for 2.5' beyond exposed edges, removed failed felts, and install new felts and gravel topping.
Deterioration due to ponding	Improper design with no, or inadequate, slopes to drains. Drains at high points on roof or obstructed.	Use additional bitumen and gravel in the low areas to limit damage. Add additional drains if possible.
Lifting at laps ("fishmouths")	Poor adhesion initially due to wrinkled felt or workmanship, or pulling as a result of blister or ridging formation.	Re-adhere if not wrinkled. Where there are wrinkles or fishmouths, cut, remove excess material, re-adhere and patch with 2 layers of felt and hot mop bitumen at least 6" beyond the laps.
Bare spots from loss of gravel	Gravel applied in adverse weather. Too thin a layer of too-fine gravel. Inadequate adhesion of gravel at edges and corners.	Broom clean, scrape off gravel, and recoat with adequate bitumen and apply a heavy dressing of properly sized gravel.
General weather deterioration of bitumen	Inadequate gravel or other surface protection, inadequate bitumen, lack of maintenance.	If felt strength affected, cannot be rejuvenated with coatings. If felts not affected, remove as much existing gravel and bitumen as possible, apply new bitumen and gravel.
Flashing failures	Rusting of metal or degradation of non-	Repair or replace flashing, seals, and

Flashing failures	Rusting of metal or degradation of non-metallic material. Inadequate allowance for movement. Inadequate fixing into reglets. Poor adhesion or inadequate protection of stripping felts. Movement at roof drains or vent pipes. Damage to capping at parapets and expansion joints. Loss of mastic or damage to mastic pans.	Repair or replace flashing, seals, and counterflashing. Do not use galvanized steel...use copper or stainless steel with corresponding fasteners. Failed or damaged boots on plumbing vents must be replaced.
Windscour	Inadequate adhesion of gravel. Ballast too small for application.	Where bare spots are due to wind scour at corners, it may be necessary to use concrete slabs as the protective cover in the area affected.
Alligator cracking	When alligator cracking of surface bitumen occurs on bare spots or on smooth-surface roofs without protective covering, adding more bitumen is only a short-term remedy; cracks will recur.	Remove alligatored material and recoat if possible. Use a felt-reinforced recoating. Protect the recoating with gravel or a paint coating.
Slippage	Slippage of gravel, felts or complete membrane on sloping roofs is due to improper choice of bitumen, thick layers of bitumen, phased construction, excessively heavy protective covering, lack of mechanical fastening where required.	If not too severe, may correct itself by change in bitumen properties on exposure. Mechanically fasten if possible, or remove and re-roof with stiffer bitumen and adequate fastening of components.
Sealing	Deterioration due to UV degradation, ozone or other atmospheric contaminants, physical abuse, normal wear and tear etc.	For skylights and similar roof construction, new gaskets, seals, or caulking may be required.

Appendix C

Interiors Preventative and Predictive Maintenance Procedures

Facility interior elements include partitions, ceilings, interior doors, built-ins, stairs, and interior finishes.

3.1.1.0—Gypsum Wallboard

Inspect interior gypsum wallboard (GWB) walls for cracks or other damage every 3 years.

WHAT TO LOOK FOR	REMEDIAL ACTIONS
Small cracks, holes smaller than 1"	Cover the crack or hole with a uniform layer of joint compound.
	Cut tape larger than the hole; center tape over crack or hole and press tape into the joint compound.
Larger cracks, holes up to 2.5"	Cover the crack or hole with a uniform layer of joint compound.
	Cut paper tape to horizontally bridge the crack or hole.
	Center tape over the crack or hole and press tape into the joint compound.
	Remove excess compound, leaving a sufficient amount under the tape to allow adhesion to wallboard.

Cover with a uniform layer of joint compound and let dry 24 hours.

Sand with a damp sponge to level surface.

Remove excess compound, leaving a sufficient amount under the tape to allow adhesion to wallboard.

Apply joint compound over the first piece of paper tape.

Place second piece of tape vertically over the first in a criss-cross pattern. Press the tape into the joint compound.

Remove excess joint compound.

Cover with a uniform layer of joint compound and let dry 24 hours.

Apply a final coat if the tape is still visible beneath the compound.

Sand with a damp sponge and blend with existing surface.

Large holes (greater than 2.5")	Cut a rectangular GWB section around the hole.
	Cut a wood or metal piece about 2" longer than the maximum opening dimension.
	Cut a GWB "patch" to fit within the newly cut opening. The patch should be about 1" larger than the opening in both directions. Trim away gypsum on the back side of the face paper, being careful not to cut the face paper.
	Screw the patch piece to the wood/metal piece, insert into opening,
	Rotate wood/metal piece to "clamp" the back of the adjacent GWB and tighten the screw to create a snug fit.
	Tape a finish the joints around the patch.

Larger areas of physical damage	Cut out damage and install new GWB.
Popped nail/screws	Install a new nail or screw 1-1/2" above the pop into the stud. When installing the fastener, you want to indent the face paper without tearing it (called "dimpling").
	Now drive and "dimple" the popped nail. Use a nail set if needed.
	Cover the slight depressions or "dimples" with joint compound and let dry 24 hours.
	Apply a second coat if your repair is highly visible and let dry approximately 24 hours.
	Sand with a damp sponge and blend with existing surface.
Torn, wrinkled face paper	Peel and remove all loose face paper.
	Coat damaged area with a primer/sealer using a paint brush.
	Apply Joint Compound to the damaged area and feather the edge to blend smoothly with the existing surface.
	Let dry approximately 24 hours.
	Apply a second coat if your repair is highly visible and let dry approximately 24 hours.
	Sand with a damp sponge and blend with existing surface.

3.1.1.1—Plaster on Lathe

Inspect interior plaster on lathe walls for cracks or other damage every 3 years. Cracks, holes, and physical damage can be repaired by removing damaged plaster and installing new.

3.1.1.2—Concrete Masonry

Inspect interior concrete masonry unit (CMU) walls for cracks or other damage every 3 years.

WHAT TO LOOK FOR	REMEDIAL ACTIONS
Horizontal crack, wall bulging, etc.	Horizontal cracks generally are more serious because they might indicate excess pressure against the wall from the opposite side or from above. Retain a Professional Engineer to evaluate condition and recommend repairs. *Do not simply "patch and paint" as problem will re-occur.*
	Once cause(s) of cracks are determined and addressed, the cracks must be repaired.
	If cracks are fine, they should not be widened. It may be adequate to paint over them with portland cement paint. Otherwise, a mortar of one part portland cement and one part sand that passes a No. 30 or No. 50 screen, depending on the size of the crack, should be worked in.
	To repair larger cracks, widen and undercut them with saw or chisel so that they are about one half inch wide at the surface, 1/2 to 3/4 inch wide at the back, and about one half inch deep. All loose dust should be brushed free. The notch should then be filled either with a dry-pack mortar or a non-sagging epoxy resin, If mortar is used, the surrounding block should be dampened thoroughly to minimize absorption of water from the mortar. A mortar of one part portland cement and two parts masonry sand should be mixed to a stiff consistency and packed in. If epoxy resin is used, it should be one that produces a gel of grease-like consistency when first mixed. When the mortar gets "thumbprint" hard, tool the joint to match existing.
	Larger cracks can also be repaired by tuckpointing. Tuckpointing is the process of removing existing mortar to a uniform depth and placing new mortar in the joint.

The existing mortar should be removed, by means of a toothing chisel or a special pointer's grinder, to a uniform depth that is twice the joint width or until sound mortar is reached. Care must be taken not to damage the brick edges. Remove all dust and debris from the joint by brushing, blowing with air or rinsing with water.

Type N, O and K mortar are generally recommended for repointing, as mortars with higher cement contents may be too strong for proper performance. The repointing mortar should be prehydrated to reduce excessive shrinkage. The proper prehydration process is as follows: All dry ingredients should be thoroughly mixed. Only enough clean water should be added to the dry mix to produce a damp consistency which will retain its shape when formed into a ball. The mortar should be mixed to this dampened condition 1 to 1-1/2 hr before adding water for placement.

The joints to be repointed should be dampened, but to ensure a good bond, the brickwork must absorb all surface water before repointing mortar is placed. Water should be added to the prehydrated mortar to bring it to a workable consistency (somewhat drier than conventional mortar). The mortar should be packed tightly into the joints in thin layers (1/4 in. maximum). The joints should be tooled to match the original profile after the last layer of mortar is "thumbprint" hard. As it may be difficult to determine which joints allow moisture to penetrate, it is advisable to repoint all mortar joints in the affected wall area.

Vertical or "stair step" crack	Most likely caused by differential stress along the base of the wall resulting from settlement of floor and footings. Retain Professional Engineer to evaluate condition and recommend repairs. Do not simply "patch and paint" as problem will re-occur. Once cause of crack is determined and addressed, the crack must be repaired...see "Horizontal Crack" above.

Crack at wall corners	Hairline cracks where the walls join other elements require caulking in accordance with Maintenance Procedure 3.3.3.4 and repainting in accordance with Maintenance Procedure 3.3.3.1.
Physical damage	Broken or cracked CMU's must be replaced. The mortar that surrounds the affected units must be cut out carefully to avoid damaging adjacent brickwork. For ease of removal, the units to be removed can be broken. Once the units are removed, all of the surrounding mortar should be carefully chiseled out, and all dust and debris should be swept out with a brush. If the units are located in the exterior wythe of a drainage wall (which is the typical case), care must be exercised to prevent debris from falling into the air space, which could block weeps and interfere with moisture drainage. The CMU surfaces in the wall should be dampened before new units are placed, but the masonry should absorb all surface moisture to ensure a good bond. The appropriate surfaces of the surrounding CMU and the replacement unit should be buttered with mortar. The replacement unit should be centered in the opening and pressed into position. The excess mortar should be removed with a trowel. Pointing around the replacement unit will help to ensure full head and bed joints. When the mortar becomes "thumbprint" hard, the joints should be tooled to match the original profile.
Failed paint	Repaint in accordance with Maintenance Procedure 3.3.3.1.

3.1.1.3—Moveable Partitions

There are three types of moveable partitions in common use, ranked in order of quality and performance life, as follows: sliding panel type, folding panel type, and accordion folding type.

Inspect moveable partitions every 6 months. During each inspection, routine maintenance is required, as follows:

1.   Clean and lubricate the track. Most tracks are aluminum...do not use abrasives or solvents to clean. Lubricate in accordance with manufacturer's instructions. Clean stains with mild detergent and

damp cloth, then buff dry with a clean cloth. Do not apply spray cleaners or polishes.

2. Adjust rollers for smooth movement with no binding or dragging.

3. Check, adjust latches.

4. Adjust sliding or folding panels, including telescopic folding panels, for alignment, hang, etc.

5. Check seals and replace as necessary.

6. Temperature and humidity will affect the operation and condition of the partitions. Check for abnormal conditions and correct.

For significant partition problems or failures, contact the manufacturer's service department for repair.

3.1.1.4—Toilet Partitions
Inspect toilet partitions annually.

Toilet partitions are made to be permanently affixed to the building. They are wall systems used primarily as privacy enclosures, which may also be mounted or anchored to the floors and ceilings of bathrooms.

All partitions

WHAT TO LOOK FOR	REMEDIAL ACTION
Anchors, fastener failure, damage	Corrosion, excess structural loading (e.g., kids climbing on partitions), etc. will cause anchors to loosen and/or fasteners to fail.
	Repair as needed using stainless steel, tamper-proof fasteners.
Structural damage	Hinges, latches, and other hardware may be damaged by vandalism. Doors may be raked or warped, or even removed. Repair as needed. Add aluminum structural members installed with stainless steel, tamper-proof fasteners to reinforce existing structural elements.
	If damage is severe, replace partitions.

Finish damage	Scratches, dings, etc. can be repaired by refinishing. See below for refinishing of hollow metal partitions. Fiberglass partitions can be repaired with gel coat patches. Stone partitions can be wet sanded.
Paper holders	Replace as necessary.

Metal partitions

WHAT TO LOOK FOR	**REMEDIAL ACTION**
Corrosion	The baked-on enamel or phenolic finish of hollow metal partitions tends to fail within 2-3 years due to abrasive cleaning methods and urine contact. Rusting of the underlying steel will then occur. If rusting is severe, replace toilet partitions. Limited rusting can be removed and the partitions repaired as follows:
	Remove rust via manual abrasion (wire brush and aluminum oxide sandpaper) and mechanical abrasion (such as an electric drill with a wire brush or a rotary whip attachment). Rust can also be removed by using a number of commercially prepared anticorrosive acid compounds. If chemicals are used, any chemical residue should be wiped off with damp cloths, then dried immediately with Industrial blow-dryers. *Do not use running water to remove chemical residue.*
	Removing rust will remove most flaking paint as well. Remaining loose or flaking paint can be removed with a chemical paint remover or with a pneumatic needle scaler or gun. (Well-bonded paint may serve to protect the metal further from corrosion and need not be removed. The paint edges should be feathered by sanding to give a good surface for repainting.)
	Once metal has been cleaned of all corrosion, small holes and uneven areas resulting from rusting should be filled with a patching material of steel fibers and an epoxy binder and sanded smooth to eliminate pockets where water can accumulate.

Bare metal should then be wiped with a cleaning solvent such as denatured alcohol, and dried immediately in preparation for the application of an anticorrosive primer (oil-alkyd based paint rich in zinc or zinc chromate), applied immediately after cleaning.

Prepare surface and repaint in accordance with Maintenance Procedure 2.2.1.11.

Once repainting is complete, install a panel of 22 ga. stainless steel on the inside surface of each partition, from the partition bottom up to 48″ above the floor level. Attach with stainless steel tamper-proof fasteners not more than 12″ on center.

3.1.1.5—Interior Glass & Storefronts

Inspect interior glass and storefronts every 3 years. See Maintenance Procedure 2.2.2.4.

3.1.2.0—Ceilings

Inspect ceilings every 3 years.

Acoustical ceilings

WHAT TO LOOK FOR	REMEDIAL ACTION
Temperature, humidity problems	Typical acoustical ceiling tiles are designed to be interior finish materials suitable for installation within a normal occupancy temperature range of 60°F to 85°F with relative humidity no higher than 70% RH.
	Ceiling materials exposed to direct contact with moisture or water, as a result of such conditions as building leaks, condensation, and HVAC system performance failure, must be replaced within 72 hours of occurrence. Remove existing ceiling materials within 72 hours, but do not install new materials until the cause of the moisture problem has been identified and corrected.

	High humidity conditions are indicated by tiles sagging or "cupping." If humidity cannot be maintained below 70% RH, replace ceiling tiles with tiles rated to 90% RH.
	Replace ceiling tiles in corridors, cafeterias, and other high humidity areas with tiles rated to 90% RH.
Dirt, dust	Use soft art gum eraser to remove small spots, dirt marks and streaks. For larger areas, use a sponge rubber pad, or wallpaper cleaner. Make sure the sponge rubber pad or wallpaper cleaner is in fresh condition. Touch up nicks and scratches with chalk. Remove dust by brushing lightly with a soft brush or clean cloth, or by vacuuming with soft brush attachment.
	If washing is required to remove dirt, do not moisten tile excessively. Never soak tile with water. Wash by light application of sponge dampened by mild liquid detergent solution: about one half capful in one gallon water. After saturating the sponge, squeeze nearly dry, then lightly rub the surface to be cleaned. Vinyl faced products are more resistant to surface moisture and can withstand repeated washings with mild detergent and anti-bacterial cleaning solutions.
Fire assembly	For many years, Underwriters Laboratories Inc. has required the use of hold-down clips in all fire resistance rated lay-in panel assemblies. Based on evaluations of data regarding gas pressures developed in fires, UL has revised their requirements as follows: *"Hold-down clips are not required for assemblies incorporating ceiling panels weighing not less than one pound per square foot."*

Gypsum wallboard ceilings

Inspect and repair in accordance with Maintenance Procedure 3.1.1.0.

3.1.3.0—Interior Doors

Inspect interior doors annually.

Frame

WHAT TO LOOK FOR	REMEDIAL ACTIONS
Frame anchored to wall	Look for signs of movement in the frame installation. Open the door about half way and attempt to rack the door. If frame moves under this test, or there is evidence of movement such as split caulking joints, wear, etc., the frame must be re-anchored. Wood and aluminum frames can be removed and reinstalled with new anchors. Hollow metal frames require new anchors and grouting to the wall.
Rusting of hollow metal	Remove light and medium rust via manual abrasion (wire brush and aluminum oxide sandpaper) and mechanical abrasion (such as an electric drill with a wire brush or a rotary whip attachment). Rust can also be removed by using a number of commercially prepared anticorrosive acid compounds. If chemicals are used, any chemical residue should be wiped off with damp cloths, then dried immediately with Industrial blow-dryers. Do not use running water to remove chemical residue. Removing rust will remove most flaking paint as well. Remaining loose or flaking paint can be removed with a chemical paint remover or with a pneumatic needle scaler or gun. (Well-bonded paint may serve to protect the metal further from corrosion and need not be removed. The paint edges should be feathered by sanding to give a good surface for repainting.) Once metal has been cleaned of all corrosion, small holes and uneven areas resulting from rusting should be filled with a patching material of steel fibers and an epoxy binder and sanded smooth to eliminate pockets where water can accumulate. Bare metal should then be wiped with a cleaning solvent such as denatured alcohol, and dried immediately in preparation for the application of an anticorrosive

	primer (oil-alkyd based paint rich in zinc or zinc chromate), applied immediately after cleaning.
Failed paint on hollow metal or wood	Prepare surface and repaint in accordance with Maintenance Procedure 2.2.1.11.
Staining, oxidation, or pitting of aluminum	Inspect aluminum surfaces for scratches or cracks in the finish. Oxidation is a natural occurrence that produces a coating that wipes off as a dark, metallic-looking residue. For optimum protection against oxidation, clean with sponge an mild soapy water and apply a coat of quality automobile wax over enamel or anodized finish on aluminum surfaces. To remove oxidation, gently scrub with fine scratch pad or steel wool; do not scratch finished surfaces, then dust or vacuum residue and wipe clean with damp cloth.
Bent or damaged frame	Hollow metal or aluminum frame sections that are bent generally cannot be straightened and must be replaced and re-anchored to the wall (see above). Damaged wood frame sections must be replaced.

Door

WHAT TO LOOK FOR	**REMEDIAL ACTIONS**
Door racked or warped, bent	Check that door fits flush against frame stops on all three sides. If not, door fit may be improved by adjusting or moving hinges. For severe racking or warp, door replacement is required.
Loose fit	Loose fit is caused by missing or failed weatherstripping, bent or damaged frame, loose hinges, etc. Check each item and repair as necessary.
Rust on metal	Remove rust via manual abrasion (wire brush and aluminum oxide sandpaper) and mechanical abrasion

(such as an electric drill with a wire brush or a rotary whip attachment). Rust can also be removed by using a number of commercially prepared anticorrosive acid compounds. If chemicals are used, any chemical residue should be wiped off with damp cloths, then dried immediately with Industrial blow-dryers. Do not use running water to remove chemical residue.

Removing rust will remove most flaking paint as well. Remaining loose or flaking paint can be removed with a chemical paint remover or with a pneumatic needle scaler or gun. (Well-bonded paint may serve to protect the metal further from corrosion and need not be removed. The paint edges should be feathered by sanding to give a good surface for repainting.)

Once metal has been cleaned of all corrosion, small holes and uneven areas resulting from rusting should be filled with a patching material of steel fibers and an epoxy binder and sanded smooth to eliminate pockets where water can accumulate.

Bare metal should then be wiped with a cleaning solvent such as denatured alcohol, and dried immediately in preparation for the application of an anticorrosive primer (oil-alkyd based paint rich in zinc or zinc chromate), applied immediately after cleaning.

Failed paint on metal or wood	Prepare surface and repaint in accordance with Maintenance Procedure 2.2.1.11.
Staining, oxidation, or pitting of aluminum	Inspect aluminum surfaces for scratches or cracks in the finish. Pay close attention to bare aluminum at edges and weepholes and areas with no finish. (Bare aluminum will oxidize rapidly in a coastal environment.) Oxidation is a natural occurrence that produces a coating that wipes off as a dark, metallic-looking residue.

For optimum protection against oxidation, clean with sponge and mild soapy water and apply a coat of quality automobile wax over enamel or anodized finish on aluminum surfaces.

	To remove oxidation, gently scrub with fine scratch pad or steel wool; do not scratch finished surfaces, then dust or vacuum residue and wipe clean with damp cloth.
Impact damage	If possible, repair damage and refinish door. If damage is severe, replace the door.

Hardware

WHAT TO LOOK FOR REMEDIAL ACTIONS

Hinges broken or bent	Replace hinges with new fasteners.
Latch, lock, or bolt not functional	Test operation and repair or replace as necessary. Check and tighten lock screws. Lubricate keyway with dry graphite lubricate as needed.
Closer/hold-open function	Closer can be adjusted for rate of swing and force of closing. Adjust force of closing so that door will close tightly against its jamb stop and the latch/lock engages. In high wind areas, adjust closer so door will not open too far.
Panic bar damaged, loose, or not functional	Test operation and repair or replace as necessary.
Stop damaged, loose, or missing	Re-anchor loose stop. Replace damaged or missing stop.
Automatic operator	See Maintenance Procedure 2.2.2.7

3.1.4.0—Specialties
Inspect specialties every 2 years. Repair or replace as required.

3.1.5.0—Firestop Systems
Firestop systems, installed correctly, require little or no routine maintenance. Routine inspections of penetrations every 5 years or so is required to ensure that building or penetration component movement has not caused separation between the substrate and the firestop systems. Where separation has occurred, the firestop system must be replaced.

Procedures must exist for managing trades people who breach any fire-resistance rated assembly while maintaining, replacing, or installing new services throughout facilities. Managers should have written procedures for firestopping and other effective compartmentation features requirements in contracts or work orders for electrical, plumbing, mechanical, cable and low-voltage contractors, building personnel, and others who might penetrate the fire- and smoke-resistance rated assemblies. This procedure should assign responsibility for verification of firestopping systems after installation. Where the quantity of penetrations is large, firestopping should be excluded from contracts or work orders and a specialty firestop contractor should be retained to complete the firestopping and compartmentation features.

3.3.3.1—Interior Painting

Inspect interior painting annually and touch-up as needed. Re-paint interior areas on the following schedule:

Corridors and other high traffic areas	Every 3 years
Classrooms	Every 5 years
Offices, conference rooms, etc.	Every 7 years
Storage rooms, mechanical rooms, etc.	Every 10 years

Interior paint must have low VOC off-gassing (maximum 50 grams/liter for flat paint and 150 grams/liter for gloss and semi-gloss) and have low toxicity, not containing any of the following heavy metals or toxic materials:

Heavy metals	Antimony, Cadmium, Hexavalent chromium, Lead, Mercury
Toxic materials	Acrolein, Formaldehyde, Naphthalene, Acrylonitrile, Isophorone, Phthalate esters, Benzene and ethylbenzene, Methyl ethyl ketone (MEK), Vinyl chloride, Butyl benzyl phthalate, Methyl isobutyl ketone, 1,1,1-trichloroethane, 1,2-dichlorobenzene, Methylene chloride, Toluene

Painting/repainting interior CMU walls
1. Painting: New or replaced CMU in dry areas shall have semi-gloss alkyd enamel finish consisting of two (2) coats of interior semi-gloss

latex enamel over one coat of latex block filler. CMU in wet areas (toilets, kitchens, janitors closets, mechanical equipment rooms, etc.) shall have polyamide epoxy coating consisting of two (2) coats of polyamide epoxy coating over one coat of latex block filler.

2.   Repainting: Prepare surface by removing loose paint, sand, and apply 1 coat of finish paint as above.

Painting/repainting interior GWB walls and ceilings
1.   Painting: New or replaced GWB shall have a semi-gloss latex enamel finish consisting of two (2) coats of interior semi-gloss latex enamel over one coat of latex-based white primer.

2.   Repainting: Prepare surface by removing loose paint, sand, and apply 1 coat of finish paint as above.

Painting/repainting acoustical ceilings:
1.   Painting: First remove all dust and dirt from surface to be repainted. Use interior acrylic or vinyl latex or alkyd paints and follow manufacturer's instructions for thinning and application. Avoid heavy brush pressure in applying paint. In spraying paint, keep spray gun pressure low and apply minimum coverage for adequate hiding. When painting, avoid clogging or bridging surface openings.

2.   Spot Painting: Spot paint badly stained areas first; then apply paint to entire ceiling area. Severe stains should be treated with a stainblocker primer/sealer before repainting to prevent bleed through.

3.   Repainting: When painting, avoid clogging or bridging surface openings. Use a paint of high hiding power since it is desirable to keep the number of coats of paint to a minimum on acoustical tile. Hiding character of paint is a particularly important consideration when a single coat is expected to cover stains or change the color of the tile. Use paint with specific formulations which have high hiding power, low combustibility and are not likely to bridge openings in the tile. Apply paint as thinly as possible.

4.   Painting/repainting fire-rated ceiling: Choose paint with extra care not to increase the Flame Spread Classification of acoustical

materials already installed. Several paint manufacturers offer formulations that are classified by Underwriters Laboratories Inc. in their Building Materials Directory. The type of paint selected or misapplication can affect the fire performance and acoustical properties of the ceiling product.

Painting/repainting concrete walls and floors

1. Clean the area: All interior masonry walls must be clean and free of loose paint, cement particles or any other foreign matter before painting. Any sources of moisture must be eliminated before painting.

2. Acid etching: New interior walls and floors should be acid etched before painting. Existing coatings can be repainted where the coating is sound. Any bare spots (including areas revealed when scraping loose and peeling paint) should be acid etched before painting.
   a. Carefully prepare a solution of one part full-strength Muriatic acid with three parts water. Always add the acid to the water to prevent the splash of hot acid. Never pour water into acid.
   b. Use one US gallon of this solution per 100 square feet of floor, and scrub with a stiff fiber bristle brush while applying. Allow the solution to remain on the floor until it stops bubbling. Flush the solution off thoroughly with clean water. If the surface is not dry within a few hours, flush it with water again. The surface must dry evenly. If puddles develop, the solution will become more concentrated. This will affect the performance of the coating applied over it.
   c. After the surface has dried, use a vacuum to remove the powder that is created by etching. Failure to remove this powder will result in poor adhesion. Painting can begin when the surface is chemically neutral and dry. When a proper etch has been attained, the concrete will have a surface texture like #1 or #2 sandpaper.

3. Apply Masonry Paint: Deteriorating concrete can produce loose sand and pebbles on the surface of a concrete wall. The most common cause for this is poor quality concrete or improper curing. Deteriorating concrete must be repaired or replaced before painting. Efflorescence is often found on the surface of interior masonry

walls and must be cleaned (see Maintenance Procedure 2.2.1.7). See Maintenance Procedure 3.3.3.1 for painting requirements.

4.   New Concrete Floors:
      a.   New concrete floors must be allowed to cure for 90 to 180 days before painting. All bare concrete floors must be acid etched before painting (see instructions for acid etching above). This is especially important where the demands of construction require the "quick-painting" of floors in less than 90 days.
      b.   Previously painted concrete floors do not need to be acid etched where paint is sound. Acid etching should be done where bare spots occur or are revealed by scraping loose or peeling paint.
      c.   Unless the floor has a moisture barrier, a latex floor coating must be used. Latex floor coatings allow moisture to pass through dry film.

3.3.3.2—Vinyl Wallcovering

Inspect vinyl wallcovering every 3 years as part of interior walls inspection.

1.   Any vinyl wallcovering showing evidence of moisture entrapment and mold growth behind it must be removed. *Do not install new vinyl in this location!* Clean wall of vinyl residue and mold, as follows, and paint wall in accordance with Maintenance Procedure 3.3.3.1.
          The two following resources are the de facto "standards" for mold mitigation and should be used to guide mold clean-up activities:
      a.   Guidelines on Assessment and Remediation of Fungi in Indoor Environments, available from the New York City Department of Health and Mental Hygiene as a free download at http://www.nyc.gov/html/doh/html/epi/moldrpt1.shtml.
      b.   Mold Remediation in Schools and Commercial Buildings, U.S. Environmental Protection Agency, EPA Publication 402-K-01-001, March 2001, available as a free download at http://www.epa.gov/mold/mold_remediation.html.

2.   If vinyl shows no evidence of mold, but has significant damage or deterioration and is more than 7 years old, replace it in its entirety as above.

3.   Minor damage or deterioration can be repaired by patching or replacing sections.

3.3.3.3—Tile

Inspect ceramic tile, terra cotta tile, etc. annually and maintain as follows:

1.   Glean grout and re-seal every 5 years.

2.   Replace cracked, broken, or missing tiles to avoid additional water problems.

Wall tiles are generally set in adhesive, while floor tiles are set in either adhesive, thinset, or mortar. Thinset is like mortar in that it is a cement-based product. It differs from true mortar in that it may be applied directly over plywood or tileboard without the need for the metal-mesh reinforcement necessary for mortar.

Replacement of floor and wall tiles is similar, as follows:

1.   Remove the grout from around the damaged tile(s) you want to replace. The grout bonds and seals the area between the tiles, protecting the floor underneath from the moisture which can eventually loosen the tiles and damage the substrate. If an attempt is made to remove the tile without removing all of the grout first, there is a chance that the adjacent tiles will chip.

   a.   With soft, unsanded wall grout, scratch it out with a utility knife with a dull blade, being careful not to slip and scratch adjacent tiles.

   b.   With hard, sanded floor grout, use a small cold chisel to get the grout out, especially if the grout line is very wide (over 1/4″). However, once the surface of the grout is broken, it may be possible to use a utility knife with the dull blade to complete the grout removal. (There is a tool called a grout saw that is intended to remove grout, but, it is useless unless the grout line is relatively wide.)

   Don't forget to save a sample of the grout for a color match!!

2.   Remove the broken tile. If the broken tile is loose, simply lift it out. For floor tiles, rap on the edge of the tile, using a hammer and a small cold chisel or old straight blade screwdriver. Do not touch any of the adjacent tile, because they may loosen or chip. A few carefully

place whacks may loosen a tile set in mortar or thinset.

If the tile is set in adhesive or well adhered to the mortar, every piece of the tile is going to fight removal. Some damage to the floor or wall underneath the tile may occur and should be patched before setting new tile.

A cold chisel or ball peen hammer can be used to break a tile into pieces, but be very careful to avoid damage to adjacent tiles. A carbide drill bit, 1/4" to 1/2" diameter, can be used to drill a series of holes in the tile, making it easier to break apart; once there is a hole in the tile, use a chisel or screwdriver to pry/break the rest of the tile out.

3.   Prepare the hole and set the replacement tile. Vacuum out all debris and scrape out any lumps or bumps in the mortar or adhesive. Test fit the new tile to make sure it sits firmly without excessive rocking and doesn't sit higher than the other tiles. Scrape out more remaining adhesive/mortar if necessary.

Apply a 1/8" layer of adhesive to the back of the tile with a putty knife or grooved tile adhesive applicator. Do not apply the adhesive closer than a half-inch to the edge of the tile.

Press the tile into its place with a slight wiggling motion, which will spread the adhesive and assure a good bond.

4.   Let the adhesive dry for 24 hours and apply matching grout. If any adhesive is squeezed out between the tiles in the last step, use a utility knife or a thin screwdriver and scrape it out. Mix the grout per instructions on the label. *Always mix no less than 2 cups of grout, regardless how little grout is actually needed, in order to get the proper mix of chemicals and pigment.* Use a damp sponge and/or fingers for pushing the grout into the cracks.

3.3.3.4—Interior Caulking

Replace interior caulking as part of painting, Maintenance Procedure 3.3.3.1; replacement of vinyl wall covering, Maintenance Procedure 3.3.3.2; partition repairs, Maintenance Procedures 3.1.1.0, 3.1.1.1, and 3.1.1.2; toilet fixture replacement, Maintenance Procedure 4.2.1.0; and/or other interior repair, replacement, or refinishing, utilizing the following sealants:

Vertical surfaces where movement is anticipated: Silicone or urethane (polyurethane). ASTM C920, Type S, Grade NS, Class 25, Use M, A, or O, as applicable.

Horizontal surfaces in high traffic areas: Urethane (polyurethane). ASTM C920, Type S or M, Grade P, Class 25, Use T. Utilize Grade NS, Use T, in areas with slopes exceeding 1 percent.

Horizontal surfaces in non-traffic areas: ASTM C920, Type S, Grade P, Class 25, Use NT. Utilize Grade NS, Use NT, in areas with slopes exceeding 1 percent.

Vertical and horizontal surfaces in humid areas (including corridors, cafeterias, kitchens, toilets, janitor closets, and mechanical equipment rooms): ASTM C920, Type S, Grade NS, Class 12-1/2, Use O.

Vertical and horizontal surfaces, dry areas, no movement anticipated: Single component water-based latex, paintable, ASTM C834.

3.3.3.5—Floor Finishes

Inspect floor finishes every 2 years and repair/replace as necessary.

Appendix D

Service Systems Preventive and Predictive Maintenance Procedures

Facility service systems include conveying systems (elevators, escalators, dumbwaiters, etc.) and plumbing, HVAC, fire protection, and electrical systems.

CONVEYING SYSTEMS

4.1.1.0—Conveying Systems

Vertical conveying systems must be inspected, tested, and maintained in accordance with the current editions of the following American Society of Mechanical Engineers (ASME) elevator codes:

ASME A17.1, "Safety Code for Elevators and Escalators": Installation requirements for new elevators, as well as maintenance and test requirements for existing equipment.

ASME A17.2, "Inspectors' Manual for Elevators and Escalators": Recommended procedures for inspection and testing of equipment to comply with A17.1 and A17.3. (A17.2.1 pertains to Electric Elevators; A17.2.2 pertains to Hydraulic Elevators)

ASME A17.3, "Safety Code for Existing Elevators and Escalators": Retroactive requirements for existing elevators and escalators. Establishes minimum standards for all elevator equipment regardless of the installation date. This code takes into account the existing building structural conditions that would limit the feasibility of bringing the elevator up to current ASME A17.1 requirements.

ASME A17.4, "Guide for Emergency Evacuation of Passengers from Elevators": Establishes procedures for the safe evacuation of passengers from stalled elevators.

The *ASME A17.1 Handbook* and *ASME A17.1 Interpretations* provide assistance for understanding and gaining insight into the code requirements and rules.

ASME QEI-1 establishes the requirements for the qualification, duties, and responsibilities of inspectors and inspection supervisors engaged in inspection and testing.

When determining the maintenance intervals, take into account the manufacturer's recommendations, how often the elevator is used, the severity of equipment loading, the age and wear of the equipment, the equipment's operating environment, and the inherent quality of the equipment.

Safety

The following practices shall be observed, at a minimum, during maintenance, inspection, or testing procedures:

1. All safety devices must be in operational condition.

2. Lockout/tagout procedures must be followed if maintenance procedures require that the equipment not be operated.

3. Ensure that personnel performing maintenance, inspection, and testing tasks wear clothing that is not loose fitting and that they are provided with proper protective equipment, such as safety shoes, hard hats, eye protection, and hand protection.

4. Provide barriers and signage, where applicable, especially at hoistway doors.

5. Upon completion of work, remove any jumper wires that were used.

6. It is possible that the elevator pit may be designated a "Permit Required Confined Space." The additional required safe procedures must be attended to in these cases.

7.　　Provide proper lighting.

8.　　Determine that adequate refuge space exists above and below the car.

9.　　Ensure the working area is clean and dry.

More detailed safety procedures can be found in publications such as the *Elevator Industry Field Employees' Safety Handbook*.

Inspections

Inspections must be performed at 6-month intervals by a Qualified Elevator Inspector, as defined by ASME QEI-1. (In some cases, such as with elevators that are rarely or never used, where the ambient conditions are good—not tending or likely to degrade and damage equipment—the recommended inspection intervals can be extended up to 12 months. Conversely, the recommended inspection interval could be shortened for some or all of the equipment, if deemed necessary.)

Specific items and equipment to be inspected under each of the following areas are listed in the Code and described in ASME A17.2.1 (for *electric* elevators) and ASME A17.2.2 (for *hydraulic* elevators), as follows:

1.　　*Inside car*: Door reopening device, stop switches, operating and control devices, car floor/landing sill, lighting, car emergency signal, car door, door closing force, power opening/closing of doors, vision panels, car enclosure, emergency exit, ventilation, signage, rated load, platform area, data plate, emergency power, restricted door opening, car ride, door monitoring, stopping accuracy.

2.　　*Machine room*: Access, head room lighting, receptacles, machine enclosure space, housekeeping, ventilation, fire suppression, pipes, wiring, ducts, guarding of equipment, numbering/labeling, disconnecting means, controller wiring/fuses/grounding, static control, overhead beam, machines and machine brakes, motor-generators, regenerated power, alternating current (AC) drives, sheaves, rope fastenings, terminal stopping devices, slack rope devices, governor, safeties, data plate.

[Note: *Hydraulic* elevators require inspection of their unique additional equipment and systems such as: heating, hydraulic

power unit, relief valves, control valve, tanks, flexible hoses/
fittings, supply line, shutoff valve, hydraulic cylinder, fluid loss
record, pressure switch, data plate, governor, recycling operation, etc.]

3. *Top of car*: Stop switch, light, outlet, operating device, refuge space,
    counterweight clearance, sheaves, normal/final terminal stopping
    devices, broken rope/chain/tape switch, leveling devices, data
    plate, emergency exit, counterweight, counterweight buffer, coun-
    terweight safeties, floor numbering, hoistway construction, smoke
    control, pipes/wiring/ducts, windows/projections/recesses/set-
    backs, clearances, multiple hoistways, traveling cables/junction
    boxes, door equipment, car frame, guide rails, guide rail alignment,
    guide rail fastenings, governor/traction/compensation ropes, rope
    fastening devices.

    [Note: *Hydraulic* elevators require inspection of their unique ad-
    ditional equipment and systems such as: terminal speed limit-
    ing devices, anti-creep limiting devices, speed test, suspension
    rope, governor rope releasing carrier, governor rope, wire rope
    fastening/hitch plate, slack rope device, traveling sheave, coun-
    terweight, etc.]

4. *Outside the hoistway*: Platform guard, hoistway doors, vision panels,
    hoistway door locking devices, access, power closing of hoistway
    doors, sequence operation, enclosure, parking devices, emergency
    access, separate counterweight hoistway, standby power selection
    switch, emergency doors in blind hoistways.

5. *Pit*: Access, lighting, stop switch, condition, clearance, runby, buf-
    fers, normal/final terminal stopping devices, traveling cables, gov-
    ernor rope, governor rope tension, compensating chains/ropes/
    sheaves, car frame/platform, car safeties, car guides.

    [Note: *Hydraulic* elevators require inspection of their unique ad-
    ditional equipment and systems such as their plunger and cyl-
    inder.]

6. Firefighters' emergency operation.

Testing

The code requires periodic testing of elevators witnessed by a Qualified Elevator Inspector. Metal test tags are required to be installed in the machine room for the Category 1 and 5 (full load) *electric* elevator tests and for the Category 1, 3, and 5 *hydraulic* elevator tests. The recommended interval for Category 1 tests is 12 months, for Category 3 tests is 36 months, and for Category 5 tests is 60 months.

Category 1 Tests—Electric Elevators: The Category 1 test requirements for *electric* elevators generally can be characterized as "no-load/ low-speed," and involve the following equipment:

1. Oil buffers
2. Safeties
3. Governor
4. Standby power operation
5. Firefighters' service
6. Door closing force
7. Final and normal stopping devices

Category 5 Tests—Electric Elevators: The Category 5 test requirements for *electric* elevators generally can be characterized as "rated-load/ rated-speed," and involve the following equipment:

1. Oil buffers
2. Safeties
3. Governor
4. Braking system
5. Emergency terminal stopping and speed-limiting devices
6. Standby power operation
7. Inner landing zone
8. Power opening of doors
9. Emergency stopping distance
10. Leveling zone and leveling speed

Category 1 Tests—Hydraulic Elevators: The Category 1 test requirements for *hydraulic* elevators involve the following equipment:

1. Relief valve setting and system pressure

2. Flexible hose and fitting
3. Hydraulic cylinder leak test
4. Standby power operation
5. Firefighters' service
6. Power operation of doors
7. Normal and final terminal stopping devices
8. Emergency terminal speed-limiting device
9. Emergency terminal stopping device
10. Pressure switch
11. Oil buffer, Safety, Governor (if provided)
12. Low oil test

Category 3 Tests—Hydraulic Elevators: The Category 3 test requirements for *hydraulic* elevators involve the following equipment:

1. Unexposed portions of pistons
2. Pressure vessels (hydrostatic test)

Category 5 Tests—Hydraulic Elevators: The Category 5 test requirements for *hydraulic* elevators involve the following equipment:

1. Oil buffer (if provided)
2. Safety (if provided)
3. Governor (if provided)
4. Coated ropes (if provided)
5. Rope fastening on pistons (if provided)
6. Overspeed valve

Pre-1970 Hydraulic Elevators

The Code did not require hydraulic elevators to possess a safety bulkhead until 1970. Hydraulic elevators installed before 1970 shall be scrutinized and frequently checked for leakage. It is recommended that these elevators have their cylinders replaced or, at minimum, that the elevator jack be provided with an external safety device to arrest the jack from uncontrolled descent.

In addition, poor insulation or lack of cathodic protection of the hydraulic cylinder may significantly shorten the life of the elevator system and render it unsafe, even if the cylinder was installed with a safety bulkhead. The use of non-corroding plastic liners is a recent design development. Most older hydraulic elevators were installed with steel liners, which are more susceptible to corrosion attack.

Prohibition of Governor Rope Lubrication
The Code prohibits the lubrication of the governor ropes.

Rope Retirement Criteria
The Code and the Inspector's Manual provide specific criteria for when the replacement of the governor, suspension, and compensation ropes are required. The ropes must be clean enough to effectively inspect them for breaks, abrasion, corrosion, wear, reduced diameter, etc.

Escalators, Moving Walks, Dumbwaiters, and Special Application Elevators
The requirements for dumbwaiters, material lifts, and special application elevators such as inclined, limited-use/limited-application, rooftop, special purpose personnel, and construction are generally the same as standard elevators and are detailed in ASME A17.1, which lists pertinent exceptions and additions.

4.1.1.1—Wheelchair Lifts
Inspect and service wheelchair lifts every 6 months as follows:

1. Check voltage and current flow (amps) under load.
2. Check bolts securing drive cabinet and base and tighten as needed.
3. Check belt tension and adjust as needed.
4. Check lift nut assembly and adjust/repair as needed.
5. Check cam rollers and replace as needed.
6. Check wear pads for excessive wear and replace as needed.
7. Inspect motor and shaft pulleys and replace as needed.
8. Check Acme screw alignment, adjust as needed.
9. Inspect and lubricate upper and lower bearings.
10. Check fastening of cable harness.
11. Check alignment of platform and doors and adjust as needed.
12. Check door interlock switch for proper operation, adjust or replace as needed.
13. Check operation of limit switch and emergency stop/alarm, adjust or replace as needed.
14. Check call/send controls at each station and on platform, adjust

or replace as needed.
15. Lubricate flip-up ramp hinge.

PLUMBING

4.2.1.0—Plumbing Fixtures
Inspect each plumbing fixture every 3 months

Tank Type Water Closets, Ballcock Flushing Mechanism

1. Inspect the base of the fixture for leaks. Remove fixture and replace seal if indicated.

2. Test looseness by grabbing the fixture and trying to rock it from side to side. If the fixture moves, tighten mounting nuts until snug, then test again. If the fixture is still loose, replace mounting bolts and/or anchors.

WHAT TO LOOK FOR	REMEDIAL ACTIONS
Water level running into top of overflow pipe	1. Remove tank top.
	2. Unscrew ball float from rod
	3. Shake ball float to determine if any water is in the ball.
	4. If water is inside the ball, replace the ball float. Otherwise, reinstall the existing ball.
	5. Place both hands on the middle of the float rod and bend the ball end of the rod down approximately 1/2."
	6. Flush WC to check that water level does not result in overflow.
	7. Replace tank top.
Damaged flush ball	1. Remove tank top.
	2. Turn off water supply at shutoff valve or main valve and flush WC to remove water.
	3. Remove flush ball from lift rod or wire and install new one.
	4. Use emery cloth or steel wool to clean flush valve outlet.

	5.	Turn on water supply and flush WC 2-3 times to make sure that new flush ball seats in flush valve outlet.
	6.	Replace tank top.
Damaged washer on ball cock assembly	1.	Remove tank top.
	2.	Turn off water supply at shutoff valve or main valve and flush WC to remove water.
	3.	Remove screws or pins holding plunger arm and plunger in place.
	4.	Lift plunger assembly.
	5.	Replace washer.
	6.	Replace plunger assembly and screws or pins holding assembly in place.
	7.	Turn on water supply and flush WC 2-3 times to make sure that new flush ball seats in flush valve outlet.
	8.	Replace tank top.
WC will not flush	1.	Remove tank top.
	2.	Check handle, trip arm, and lift rod. Adjust or replace as needed.
	3.	Flush WC 2-3 times to make sure that new flush ball seats in flush valve outlet.
	4.	Replace tank top.

Tank Type Water Closets, Float Cup Flushing Mechanism:

1. Inspect the base of the fixture for leaks. Remove fixture and replace seal if indicated.

2. Test looseness by grabbing the fixture and trying to rock it from side to side. If the fixture moves, tighten mounting nuts until snug, then test again. If the fixture is still loose, replace mounting bolts and/or anchors.

WHAT TO LOOK FOR	REMEDIAL ACTIONS
Water level running into top of overflow pipe	1. Remove tank top.
	2. Grab the top and bottom of the adjustment clip, squeeze and move it down the pull rod to lower the float cup.

	3. Flush WC to check that water level does not result in overflow.
	4. Replace tank top.
Running water closet	1. Remove tank top.
	2. Turn off water supply at shutoff valve or main valve and flush WC to remove water.
	3. Lift flapper ball and check for damage or wear, replace as needed.
	4. Use emery cloth or steel wool to clean flush valve outlet.
	5. Turn on water supply and flush WC 2-3 times to make sure that new flush ball seats in flush valve outlet.
	6. Replace tank top.
WC will not flush	1. Remove tank top.
	2. Check handle, trip arm, and chain. Adjust or replace as needed.
	3. Flush WC 2-3 times to make sure that new flush ball seats in flush valve outlet.
	4. Replace tank top.

Flush Valve Water Closets and Urinals

1. Inspect the base of the fixture for leaks. Remove fixture and replace seal if indicated.

2. Test looseness by grabbing the fixture and trying to rock it from side to side. If the fixture moves, tighten mounting nuts until snug, then test again. If the fixture is still loose, replace mounting bolts and/or anchors.

3. Test flush valves for proper operation and repair or replace as follows:

WHAT TO LOOK FOR	PROBABLE CAUSE	REMEDIAL ACTIONS
Flush does not function (no flush)	Stop valve or zone valve closed.	Check and open valve(s)
	Handle assembly is damaged.	Replace handle or repair with handle repair kit.

	Relief valve is damaged.	Replace relief valve.
Handle leaks	Handle seal or handle assembly is damaged.	Replace handle or repair handle repair kit.
Water splashes from fixture	Water pressure is too high (max 80 psig)	Use stop valve to reduce pressure (and flow).
	Conventional diaphragm is installed on a low-flow fixture.	Determine required fixture flush volume and install correct diaphragm assembly or relief valve.
Flow volume does not adequately siphon fixture	Stop valve or zone valve partially closed.	Check and open valve(s).
	Diaphragm assembly is damaged.	Replace diaphragm assembly.
	Conventional diaphragm is installed on a low-flow fixture.	Determine required fixture flush volume and install correct diaphragm assembly or relief valve.
	Inadequate water pressure or flow.	Determine maximum water pressure available; must be at least 40 psig.
		If pressure is adequate, install a higher flushing volume relief valve and/or diaphragm assembly.
Flush valve closes too quickly (short flush)	Worn or damaged diaphragm assembly.	Replace diaphragm assembly.
	Handle assembly is damaged.	Replace handle or repair with handle repair kit.
	Conventional diaphragm is installed on a low-flow fixture.	Determine required fixture flush volume and install correct diaphragm assembly or relief valve.

Flush valves stays open too long (long flush) or fails to shut off.	Bypass hole of the diaphragm assembly is clogged.	Remove the diaphragm assembly and disassemble the filter rings. Rinse under running water and reassemble.
	Relief valve or diaphragm assembly is worn or damaged.	Replace relief valve or diaphragm assembly.
	Relief valve is not seated properly.	Remove the diaphragm assembly and disassemble components. Rinse under running water and reassemble.
	Conventional diaphragm is installed on a low-flow fixture.	Determine required fixture flush volume and install correct diaphragm assembly or relief valve.
Chattering noise during flush	Inside cover is damaged.	Install new cover.
	Relief valve or diaphragm assembly is worn or damaged.	Replace relief valve or diaphragm assembly.

Sinks and Lavatories:

1. Sinks may have washerless faucets or standard compression faucets with a washer. If a faucet leaks or malfunctions, repair or replace as indicated.

2. Clean faucet aerators. [In high risk applications (prisons, hospitals, nursing homes, etc.), remove aerators each month to remove sediment and scale and clean to prevent *Legionella*.] To clean an aerator, unscrew it from the mouth of the faucet, remove any deposits, remove and rinse the washers and screens with chlorine bleach, replace in their original order and put back on the faucet.

3. Inspect for damage, cracked or broken ceramic, etc. Test mounting for looseness and tighten as required.

Showers

Inspect shower valves and heads, drains, and enclosure, as follows:

WHAT TO LOOK FOR	PROBABLE CAUSE	REMEDIAL ACTIONS
Cracked or damaged ceramic tile	Vandalism or age.	Repair/replace in accordance with Maintenance Procedure 3.3.3.3.
Drain cover damaged or missing, drain stopped-up	Vandalism or age.	Clean and repair as necessary.
Sewer gas or odors from drain	Trap has dried-out.	Fill trap with water (turn on shower for 30 seconds).
Shower head	In high risk applications (prisons, hospitals, nursing homes, etc.), remove each month to remove sediment and scale and clean with chlorine bleach to prevent *Legionella*.	
Shower valve condition and operation	Test maximum supply water temperature. If it exceeds desired temperature (125°F, unless a lower temperature is required by code or regulation), adjust valve to reduce supply temperature. Service shower valves annually. Utilize the manufacturer's "service kit" and replace all perishable parts in accordance with the manufacturer's instructions. After disassembly, soak all metal parts within the valve in descalent and wash off with clean water. Use silicone-based grease on all seals and moving parts to ensure smooth operation and re-assemble.	
Shower head condition and operation	Spray pattern weak or broken.	Disassemble and clean head.
	Head damaged, arm bent or broken, escutcheon missing, etc.	Replace.

Vanity curtain and hardware condition	Dirty or damaged curtain, rings broken or missing, etc.	Replace.
	Hardware damaged or worn.	Repair or replace as required.

Drinking Fountains/Electric Water Coolers

WHAT TO LOOK FOR	PROBABLE CAUSE	REMEDIAL ACTIONS
Water is too warm	Compressor not running due to overload/relay, broken or loose wire, no power.	Repair. If overload has tripped, check condenser (see below) for high head condition.
	Compressor failure.	Check compressor, replace if needed.
	Compressor low on refrigerant.	Find and repair leak. Recharge system.
	Temperature control set too cold or defective.	Reset or replace as needed.
	Condenser dirty, fan motor failure, fan failure.	Clean condenser, replace motor and/or fan blade.
Water is too cold	Temperature control set too cold or defective.	Reset or replace as needed.
No water flow from bubbler or erratic water flow from bubbler	Water supply flow or pressure problem.	Supply pressure should be 40-80 psig.
	No electrical power (solenoid closed, compressor not running).	Check power, restore.
	Clogged inlet strainer.	Remove strainer and clean. Replace if necessary.

	Water tank freezing (temperature control set too cold or defective).	Reset or replace as needed.
Water "trickles" out of bubbler	Actuator/push bar adjustment.	Service and align actuator for proper operation.
	Debris in regulator.	Disassemble regulator and clean.
	Temperature control set too cold or defective.	Reset or replace as needed.
Water flows continuously from bubbler	Debris in regulator or water in the regulator housing.	Disassemble regulator and clean.
	Worn or defective regulator.	Replace.
Water stream too high or too low	Water pressure too high or too low.	Supply pressure should be 40-80 psig.
	Worn or defective regulator.	Adjust regulator for desired stream height. If not possible, replace regulator.
	Clogged inlet strainer.	Remove strainer and clean. Replace if necessary.
Compressor runs excessively	Temperature control set too cold or defective.	Reset or replace as needed.
	Compressor low on refrigerant.	Find and repair leak. Re-charge system.
	Condenser dirty, fan motor failure, fan failure.	Clean condenser, replace motor and/or fan blade.
Unit operation is noisy	Compressor mounting loose or failed.	Repair or replace.
	Condenser fan blade hitting condenser or housing.	Replace fan and/or motor.

	Condenser fan motor bearings.	Replace fan motor.
	Tubing or casing rattle.	Find rattle and repair.
Casing corrosion or damage	Age or vandalism	If minor, repair. If significant, replace drinking fountain.

Caulking

Inspect caulking around fixtures, looking for signs of deterioration or damage. Remove the old caulking and replace with new as needed.

4.2.1.1—Kitchen Sinks

Inspect and service kitchen sinks annually, as follows:

1. Sinks may have washerless faucets or standard compression faucets with a washer. If a faucet leaks or malfunctions, repair or replace as indicated.

2. Clean faucet aerators. [In high risk applications (prisons, hospitals, nursing homes, etc.), remove aerators each month to remove sediment and scale and clean to prevent *Legionella*.] To clean an aerator, unscrew it from the mouth of the faucet, remove any deposits, remove and rinse the washers and screens with chlorine bleach, replace in their original order and put back on the faucet.

3. Inspect for damage, cracked or broken ceramic, etc. Test mounting for looseness and tighten as required.

4.2.1.2—Dishwashers

Inspect and service dishwashers annually in accordance with manufacturer's instructions.

4.2.1.3—Grease Traps & Art Room Traps

Inspect grease traps and art room traps monthly. Clean as required.

4.2.1.4—Water Heaters

Inspect and service water heaters annually, as follows:

1. To prevent *Legionella*, maintain hot water storage temperature at

140°F or higher—the hotter the better. Hot water supply temperature should be at least 125°F unless codes or regulations dictate a lower temperature to prevent scalding. In high risk applications (prisons, hospitals, nursing homes, etc.), consider the use of copper-silver ionization water treatment.

2. Test pressure/temperature relief valve for proper operation. Lift or depress the lever and drain water from the overflow pipe. If water doesn't drain out, shut off water to the heater, open a hot water faucet somewhere, and replace the valve.

3. Check controls for proper hot water supply temperature. Test by opening a hot water faucet for three minutes. If the heater doesn't turn on, reset the control to a lower temperature and test again. If it still fails, replace temperature controller.

    If heater operates, but water temperature is too low:
    a. Test electric heater for defective element(s), defective thermostat, and/or defective safety thermostat. Replace as required. [Note: In hard water areas, heater elements have scale deposits that reduce heat transfer. Inspect, and if necessary, remove elements and soak in vinegar, then scrape to remove scale.]

    b. For gas or oil-fired heater, test burner for proper firing. Check for defective thermocouple. Repair/replace as required.

    c. For oil-fired heater, check for clogged or dirty burner. Clean as required.

    d. For steam heaters, check operation of control valves. Repair valves, replace operators as required.

    If heater operates, but water temperature is too high:
    a. Test electric heater for defective thermostat and/or defective safety thermostat. Replace as required.

    b. For gas or oil-fired heater, test burner for proper firing. Check for defective thermostat or exhaust vent is restricted. Repair/replace as required.

    c. For steam heaters, check operation of control valves. Repair valves, replace operators as required.

4.   Inspect for leaks, corrosion, etc. and repair as needed.

5.   Open the drain valve at the bottom of tank type heater, letting the water run into a bucket until it looks clear (usually about five gallons) to prevent sediment accumulation.

6.   Inspect and service gas or oil burners in accordance with Maintenance Procedure 4.3.4.4.

7.   Inspect the flue assembly for leaks. Replace any corroded vent piping, seal joints, adjust hangers, etc.

4.2.2.0—Service Water Piping

Inspect service water piping systems as follows:

1.   Inspect annually for leaks.

2.   Inspect and service water meter annually, as follows:
    a. Remove meter head from the line and check the mechanism and the condition of the line and straightening vanes—check for clogged or obstructed line or vanes.
    b. Check for water accumulation inside the meter. Replace seals as required.
    c. Check and clean all meter parts in the flow stream. Make sure all moving parts spin freely.
    d. Check front bearing for excessive "play"—replace if required.
    e. Lubricate register clock with light-weight oil. Grease all other components fitted with grease fittings. *Do not over lubricate!*
    f. Check meter indicator for proper operation—replace indicator is gears and bushings are worn or bound.

2.   Every 5 years, inspect to evaluate condition. Adjust anticipated performance life in accordance with current condition.

3.   Every 5 years, if water piping was installed before 1986, take water samples in at least 3 locations and send to certified lab for lead testing. If lead is found, immediately implement short term lead avoidance plan. Reduce anticipated performance life of piping system to 3 years for planned replacement.

4.2.2.1—Piping Systems

Inspect piping systems annually for leaks and every 5 years inspect to evaluate condition. Adjust anticipated performance life in accordance with current condition.

4.2.2.2—Backflow Preventers

The following are annual tests and procedures for inspecting, testing, and repairing a reduced pressure principle (RPP) or reduced pressure zone (RPZ) type backflow preventer, based on the following test setup:

Tests Set-Up

1. Backflow preventers are typically provided with four (4) pressure taps—one upstream of the inlet isolation valve (test cock No. 1), one between the inlet isolation valve and Check Valve No. 1 (test cock No. 2), one located in the reduced pressure zone (test cock No. 3), and one located downstream of Check Valve No. 2 (test cock No. 4).

2. Connect the high pressure port of a differential pressure gauge (0-15 psi) to test cock No. 2 and the low pressure port of the gauge to test cock No. 3.

Preliminary Investigation and Evaluation

1. Look for signs of leakage from the relief port valve.

2. Inspect for conditions which could prevent normal functioning of the device, plugged relief valve port, etc.

3. Determine if the device has been properly installed above ground or floor level and is protected from freezing.

4. Determine if any discharge from the relief valve port would be visible and that the port is not directly connected to a sewer.

Test No. 1—Relief Valve Opening Point

Test the operation of the differential pressure relief valve. The differential pressure relief valve must operate to maintain the zone between the two check valves at least 2 psi less than the pressure of the supply side of Check Valve No. 1. Perform the following procedure:

1. Bleed the test cocks in the following order. First open test cock No. 4 and leave it open while bleeding each of the other test cocks individually starting with No.1, then No.2 and the No. 3. Open each of these three test cocks slowly and then close before proceeding to the next one. After test cocks 1 through 3 have been flushed and shut off, then close test cock No. 4.

2. Install appropriate fittings to attach gauge hoses to test cocks No. 2, 3 and 4.

3. Attach hose from high side of the differential pressure gauge to the No. 2 test cock.

4. Attach hose from low side of the differential pressure gauge to the No. 3 test cock.

5. Open test cock No. 3 slowly and then bleed all air from the hose and gauge by opening the low side bleed needle valve.

6. Leaving the low side bleed needle valve open, slowly open test cock No. 2 and then bleed all air from the hose and gauge by opening the high side bleed valve.

7. Close the high side bleed needle valve after all air is expelled and then slowly close the low side needle valve.

8. Close the No. 2 Shutoff Valve and note the position of the needle on the differential pressure gauge. If the needle continues to drop, then the No. 1 Check Valve is leaking and the rest of the testing cannot be completed. If the needle remains steady, then note its position as the differential pressure drop across the No. 1 Check Valve.

9. Open the high side control needle valve approximately one turn and then open the low side needle control valve no more than a quarter of a turn so that the differential gauge needle drops slowly. Observe the opening point of the relief valve by placing your hand where the water will drip on it and record the gauge reading when the relief valve first drips.

10. Close the low side needle control valve.

Test No. 2—Tightness of the No. 2 Check Valve

Test the No. 2 Check Valve for tightness against backpressure. Perform the following procedure:

1. Maintain the No. 2 Shutoff Valve closed from the first test and the high side control needle valve open.

2. Vent all of the air through the bypass hose by opening the bypass needle valve.

3. With the bypass hose venting a small amount of water, attach it to the No. 4 test cock and then close the bypass needle valve. After the bypass needle valve is closed, open the No. 4 test cock.

4. Bleed water from the zone by opening the low side bleed valve on the gauge to re-establish the normal reduced pressure within the zone. Once the gauge needle reaches a value above the noted No. 1 Check Valve pressure drop (step 8 of Test No. 1), close the low side bleed valve.

5. Open the bypass needle valve and observe the position of the needle on the gauge. If the indicated differential pressure reading remains steady, then the No. 2 Check Valve is reported as "Closed Tight." Go to Test No. 3.

    If the differential pressure reading falls to the relief valve opening point, bleed water through the low side bleed needle valve until the gauge reaches a value above the noted No. 1 Check Valve pressure drop. If the gauge needle settles above the relief valve opening point, record the No. 2 Check Valve as "Closed Tight" and proceed to test No. 3. If the differential pressure gauge reading falls to the relief valve opening point again, then the No. 2 Check Valve is reported as "leaking" and Test No. 3 below cannot be completed.

    If the differential pressure reading drops, but stabilizes above the relief valve opening point, the No. 2 Check Valve can still be reported as "Closed Tight."

    If the gauge needle continues to rise, then a check for backpressure must be conducted and the situation corrected before testing can be completed.

Test No. 3—Tightness of No. 1 Check Valve

Determine the tightness of Check Valve No. 1, and to record the static pressure drop across Check Valve No. 1. The static pressure drop across Check Valve No. 1 should be at least 3.0 psi greater than the relief valve opening point (see test No. 1). This 3.0 buffer will prevent the relief valve from discharging during small fluctuations in line pressure. A buffer of less than 3.0 psi does not imply a leaking Check Valve No. 1, but rather is an indication of how well it is sealing. Perform the following procedure:

1.   With the bypass hose connected to test cock No. 4 as in Step 3 of Test No. 2 (high side control needle valve and bypass needle valve remaining open), bleed water from the zone through the low side bleed needle valve on the gauge until the gauge reading exceeds the noted No. 1 Check Valve pressure drop. Close the low side bleed needle valve. After the gauge reading settles, the reading is the actual static pressure drop across Check Valve No. 1 and should be recorded as such.

2.   Close all test cocks on assembly and slowly open Shutoff Valve No. 2 returning assembly to service. Open high side and low side bleed valves to drain gauge and remove all hoses. Open all needle valves on gauge and drain water from the gauge.

Troubleshooting

The following troubleshooting guide can be use to diagnose and repair pressure reducing valve problems:

WHAT TO LOOK FOR	PROBABLE CAUSES	REMEDIAL ACTIONS
Relief valve continuously discharges during no-flow condition (only).	No. 1 check valve fouled or not moving freely or fNo. 2 check valve fouled, coupled with a backpressure condition. disc	Inspect and clean seat disc and seats of check valves.
Relief valve continuously discharges at all times (flow and no-flow conditions).	a. Relief valve fouled. b. Damaged diaphragm. c. Sensing tube to inlet side of diaphragm plugged.	a. Inspect and clean relief valve seat disc and seat. b. Replace diaphragm. c. Inspect and clean tube. d. Inspect and clean seat

	d. No. 1 check valve not moving freely	disc and seat of check valve.
Relief valve discharges intermittently during no-flow condition	Pressure fluctuations on water supply side.	Find cause of pressure fluctuations and eliminate them.
Relief valve does not open during Test No. 1	a. No. 2 shut-off valve not closed completely. b. Test equipment improperly installed.	a. Close No. 2 shut-off valve. b. Recheck test procedures.
No. 2 check valve fails to hold backpressure	a. No. 2 shut-off valve not closed completely. b. No. 2 check valve fouled. c. No. 1 check valve not moving freely.	a. Close No. 2 shut-off valve. b. Inspect and clean seat disc and seat of No. 2 check valve. c. Inspect and clean seat disc and seat of No. 1 check valve.
Pressure differential across No. 1 check valve is low during Test No. 3	a. No. 1 check valve fouled or not moving freely. b. Pressure fluctuations on water supply side causing inaccurate gauge reading.	a. Inspect and clean seat disc and seat of No. 1 check valve. b. Find cause of pressure fluctuations and eliminate them.

4.2.2.3—*Legionella*

To monitor and control *Legionella* in plumbing systems, the following maintenance measures are recommended in ASHRAE Guideline 12-2000, *Minimizing the Risk of Legionellosis Associated with Building Water Systems.*

1. Hot water should be stored at temperatures of 140°F+—the hotter the better.

2. Elevated holding tanks for hot and cold water should be inspected and cleaned annually.

3. Copper-silver ionization water treatment should be used for high-risk applications such as prisons, hospitals, nursing homes, etc..

4. In high-risk applications, showerheads and faucet aerators should be removed monthly to remove sediment and scale and to clean them in chlorine bleach.

5.   Emergency shower and eyewash stations should be flushed at least monthly.

6.   Appropriate precautions should be taken when testing any fire protection system.

7.   High temperature flushing or chlorination is recommended if Legionella is found in service water systems.

There are questions relative to the need and/or effectiveness of some of the measures listed above. Measures implemented in the U.K. and Netherlands, and even in some locales in the United States, vary significantly from these recommendations, yet have proven quite effective. To date, there is little research and few studies on this topic. *Therefore, maintenance directors, especially those in high risk facilities, must stay abreast of this debate and evaluate new maintenance measures as they are presented.*

Hot Water Systems
    It is recommended that water heaters be kept at a minimum of 140°F and all water be delivered at each outlet at a minimum of 125°F to reduce the potential for Legionella development. (For facilities that are required to maintain water temperatures to prevent scalding should be equipped with thermostatically-controlled mixing valves at each shower/bath to maintain lower water temperatures.)
    It is essential to identify all parts of the domestic water systems where water may stagnate (e.g., "dead legs" or laterals that have been capped off, storage tanks that have "dead zones" or are not frequently used). For treatment to be effective, the stagnant zones must be removed from the system. Rubber and plastic gaskets in the plumbing system may also serve as a *Legionella* growth medium. Eliminate or minimize use of these materials and substitute materials not conducive to *Legionella* growth. Follow Procedure 4.2.2.2 to test the integrity of all backflow preventers.

To test for Legionella
1.   Collect water samples before beginning treatment to determine potential contamination. Draw 200 milliliters to 1 liter of water from the draw-off valve of all water heaters into a sterile container. Check the temperature of the water in these units to determine if it is sig-

nificantly lower than the set temperature. Sample a representative number of domestic hot-water faucets or outlets. It is important not to flush the faucet before taking a sample because the end section of the water system may be a source of contamination. Collect a 200 milliliter to 1 liter "preflush sample" of the first hot water drawn from the outlet. Allow the water to run and measure the temperature, and then collect a second, "postflush" sample when the water temperature is constant.

2. Submit the water samples to a laboratory qualified to measure CFU of *Legionella* per milliliter of water using the CDC "Standard Culture Method."

If *Legionella* is detected, use the clean-up procedure below to treat all hot-water systems that have either been tested and found to contain detectable levels of *Legionella* or have been assumed to be contaminated:

1. Disinfect the system using any effective chemical, thermal, or other treatment method. For example:
    a. Pasteurize the hot water system by heating the water to at least 160°F and maintain this temperature for a minimum of 24 hours. While maintaining the temperature at 160°F, continuously flush each faucet on the system with hot water for 20 minutes.
    b. Use an accepted chemical disinfectant such as chlorine or an acceptable biocide treatment to clean the system. Thoroughly flush the system after treatment to remove all traces of the corrosive and possibly toxic chemicals.
    c. Follow any other technique that has demonstrated effectiveness and safety.

2. After treatment, resample the hot water from each storage tank. If *Legionella* are detected, re-treat and resample the water system. If no measurable levels are found in this system and all other potential sources have also been addressed, go to the next step.

3. Test the domestic hot- or warm-water system for *Legionella* on the following schedule to assure that recontamination has not occurred:
    a. Weekly for the first month after resumption of operation.
    b. Every two weeks for the next two months.

c.   Monthly for the next three months.

4.   During the monitoring period, if 10 or more CFU per milliliter of
     water are present, re-treat the system according to steps 1-2 above.
     Resume weekly testing (step 3a) after retreatment. If levels remain
     below 1 CFU per milliliter, no further monitoring is necessary. If
     the levels are between 1 and 9 CFU per milliliter, continue monthly
     sampling of the water indefinitely and continue efforts to deter-
     mine the source of contamination. Make test results available to
     building residents.

Tepid Water Systems

**Warm-water systems or tepid water systems, usually associated
with eyewash safety systems, dilute domestic hot water from a water
heater with cold water upstream from the outlet source are not recommended. Warm water left in these lines is at ideal temperatures
for amplification of *L. pneumophila*. Localized mixing at the source
to temper very hot water is more acceptable. Another alternative is
"instantaneous" point of delivery heating of water using individual
steam heating systems at each outlet.**

Cold-Water Systems

Domestic cold-water systems have not been a major source of concern for Legionnaires' disease because *L. pneumophila* will not amplify
at low temperatures. Cold-water storage and delivery should be at less
than 65°F to minimize potential for growth. cold-water lines near hot-
water lines should be insulated. Try to eliminate stagnant places in the
system as dead legs or storage tanks that are not routinely used.

Detectable levels of *L. pneumophila* in the system may indicate contamination of the source water supply and should represent the maximum allowable level in the system. If sampling of the system indicates
a level of contamination significantly greater than that of the incoming
domestic water supply system, treat the system and identify the source
of contamination or amplification. Follow the same clean-up procedure
listed for hot water systems above if cold-water systems are shown to
contain measurable *Legionella* or are assumed to be contaminated.

4.2.3.0—Floor Drains

Inspect floor drains every 4 months, as follows:

WHAT TO LOOK FOR	PROBABLE CAUSE	REMEDIAL ACTIONS
Cracked or damaged ceramic tile	Vandalism or age.	Repair/replace in accordance with Maintenance Procedure 3.3.3.3.
Drain damaged or stopped-up	Vandalism or age	Clean and repair as necessary.
Sewer gas or odors from drain	Trap has dried-out.	Fill trap with water. (Note: for locations that are seldom wet, fill trap with mineral oil to reduce evaporation.)

4.2.4.0—Gas Piping
Inspect gas piping annually, as follows:

1. Using a *combustible gas detector*, check every joint and connection in exposed piping. (There should be no concealed piping. If concealed piping is found, replace or reroute to make piping exposed.)

2. If underground piping is installed, have a pressure test to 1.5 times the maximum working pressure or at least 3 psig performed every 5 years.

3. Check that the main gas shutoff valve and the gas shutoff valve at each appliance functions properly. If valve does not function, replace.

4. Check that a drip leg is installed at each appliance. If not, install one.

5. Check condition of flexible gas lines and connections for leaks or damage. Replace flexible copper lines with flexible stainless steel lines.

6. Check lines for proper support. Add supports, hangers, etc. as required.

7. Check appliance flame condition. If flame is yellow or orange, "lazy," unusually high, makes a "popping" sound when turned on

or off, or burns at the orifice, appliance may require repair. *Valve off and disconnect gas supply until appliance repairs are made.*

8.   If gas is supplied from an LP tank, the following is required
      a.   Check that tank is secure and level. Adjust and anchor as needed.
      b.   Examine tank for corrosion or other damage. Replace as required.
      c.   Make sure that relief safety valve and primary pressure regulator are protected from icing.
      d.   Have tank vendor inspect tank on a routine basis and test primary pressure regulator at least once a year.

4.2.4.1—Acid Waste Systems
Inspect and service acid waste systems, as follows:

Dilution neutralization
1.   Each month, check neutralization system operation, including test of controls, pH sensors, agitator, chemical feed pump, and solenoid valves. Repair or replace components as required.

2.   Every two years, have pipe tested for leaks and corrosion. Replace piping sections as required with polypropylene piping using heat fused/welded or mechanical joints. *Do not use air to test—use water test as specified below.*

Limestone/marble chips neutralization
1.   Each month, test effluent for proper pH. Add limestone/marble chips as required to maintain discharge pH between 7.2 and 7.8.

2.   Every two years, have pipe tested for leaks and corrosion. Replace piping sections as required with polypropylene piping using heat fused/welded or mechanical joints. *Do not use air to test—use water test as specified below.*

Water leak test
To test pipe, insert plugs to isolate piping section to be tested. All openings to that section must be plugged or capped with test plugs or caps. Then, fill the system with water to the highest point in the system.

Filling the system slowly should allow air in the system to escape. Any air trapped in the system must be expelled before beginning the test. Maintain water in the system, with no change in level, for four (4) hours.

4.2.4.2—Pool Maintenance

Swimming or wading pool maintenance consists of routine water chemistry testing and adjustment, along with routine inspection and servicing of pool piping and equipment, as follows:

1. Test and adjust water chemistry daily. A pool that is "balanced" has proper levels of pH, Total Alkalinity and Calcium Hardness. It may also be defined as water that is neither corrosive or scaling. Using a good test kit with fresh testing reagents, measure the chemical parameters of pH, alkalinity and calcium hardness and adjust as necessary

    a. To have pH in balance, adjust the water with additions of a pH increaser (base) or a pH decreaser (acid) to achieve the range of 7.2 - 7.8. If testing shows a pH value below 7.2, the water is in an corrosive (acidic) condition and sodium carbonate must be added to bring the pH into a more basic range and prevent corrosion. Conversely, if the pH is above 7.8, the water is in a scaling (basic) condition and Muriatic acid must be added to bring down the pH.

    b. Low alkalinity is raised by the addition of sodium bicarbonate. High levels of alkalinity are lowered by the addition of Muriatic acid. (Experts recommend "pooling" the acid in a small area of low current for a greater effect on alkalinity. That is, adding an acid will lower both pH and alkalinity. Walking the acid around the pool, in a highly distributed manner is said to have a greater effect lowering the pH than the alkalinity. Pooling the acid has the opposite effect.) Alkalinity should be maintained within 80-120 ppm.

    c. The test for Calcium Hardness is a measure of how hard or soft the water is. If Calcium Hardness levels are too high, add TSP and/or add fresh water to lower the levels. Levels that are too low require the addition of calcium chloride. Recommended range for calcium hardness is 200 - 400 ppm.

    d. The Saturation Index, also called the Langelier Index, is a chemical equation used to diagnose the water balance in the pool. To calculate the Saturation Index, measure pH, temperature, calcium hardness, and total alkalinity. Refer to chart or pool calculator for assigned values corresponding to the temperature, hardness, and alkalinity readings and add these to your pH value. Subtract 12.1, which is the constant value assigned to Total Dissolved Solids, and a resultant number will be produced. A result between -0.3 and +0.5 is said to indicate balanced water. Results outside of these parameters require adjustment to one or more chemical components to achieve balance, within the individual component ranges specified above.

2.   Daily, test and adjust chlorine level (sanitizer and anti-algae treatment) to maintain the recommended level of free available chlorine between 1.0 and 3.0 ppm.

3.   Weekly, inspect diving boards, ladders, ropes, etc. to ensure they are not damaged, loose, or otherwise unsafe. Check pool decks for broken or cracked tiles or other surface damage that must be repaired.

4.   Inspect pool piping and equipment weekly
      a.   Clean out the pool's skimmer basket.
      b.   Open the pump strainer basket and clean it.
      c.   Check the pressure drop across the filter. (There is no point in checking it before cleaning out the skimmer and strainer baskets; if they are full, the filter pressure will be low). If the pressure is high, the filter needs backwashing.
      d.   Check the heater. Turn the heater on and off a few times to make sure it is operating properly. While the heater is running, turn the pump off. The heater should shut off by itself when the pressure from the pump drops.
      e.   Check the time clocks for the correct time of day and the settings for the time(s) for the daily filter runs, etc. (Trippers come loose and power fluctuations or unrelated service work can affect the clocks.)
      f.   Look for leaks or other signs of equipment failure. Clean up the equipment area by removing any debris from around the motor vents and heater. If the pool equipment is exposed, clear drains of debris that could prevent water from draining away from the equipment during rain.

HEATING, VENTILATING, AND AIR-CONDITIONING

WARNING! **When performing maintenance of any motor-driven HVAC equipment, an electrical "lock out/tag out" procedure may be required to prevent accidental starting of the equipment. This procedure must establish minimum requirements for controlling hazardous energy whenever maintenance or repair is done on machinery, equipment, and property. It is used to ensure that the machine or equipment is stopped, isolated from all potentially hazardous energy sources and locked out before employees perform any servicing or maintenance where the unexpected energizing or start-up of the machine or equipment or release of stored energy could cause injury.**

4.3.1.0—Packaged Incremental HVAC Units

Inspect packaged incremental HVAC units annually at the end of the cooling season. If unit is over 5 years old and shows significant rust or other physical damage or deterioration or has performance problems, replace.

4.3.1.1—Split System Units

Inspect and service packaged or split system air-conditioners and heat pumps.

Inspect and service the airside section of each packaged unit or the indoor unit of a split system in accordance with Maintenance Procedure 4.3.3.2.

Inspect the compressor/condensing section of each packaged unit or the condensing/outdoor unit of each split system every 6 months. Service as follows:

1. Unit casing: Check and clear any leaf litter or organic matter in contact with the casing. Remove leaves, sticks, etc on or in the unit casing. Check for condition of paint, metal, etc. and repair as necessary.

2. High and Low Pressure Switches: Switches have fixed, nonadjustable settings. Check operation by slowly closing the liquid shutoff valve and allowing compressor to pump-down below 1 psig. Compressor should shut down when suction pressure drops to cut-out pressure (see manufacturer's data) and should restart when pressure builds up to cut-in pressure shown.

3. Outdoor Fans: Check for proper rotation and that fans do not run backwards when off. Clean and, if needed, balance fan blades.

4. Lubrication: Typically fan motors have sealed bearings and require no lubrication. (However, if motor does require lubrication, do so in accordance with Maintenance Procedure 4.3.20.0.)

5. Coils: Clean coils in accordance with Maintenance Procedure 4.3.3.3.

6. Check refrigerant charge
    a. Insert a digital thermometer under the insulation on the liquid line near the filter-dyer.
    b. Connect pressure gauge to the compressor discharge line.
    c. Operate unit for 15 minutes and read discharge (head) pressure.
    d. Using "Pressure-Temperature" chart for the refrigerant utilized, find the equivalent saturated condensing temperature.
    e. Read the liquid line temperature and subtract the saturated condensing temperature determined in Step d. This difference is the sub-cooling temperature.
    f. Compare this temperature to the normal sub-cooling temperature listed in the manufacturer's service literature for the unit. If the difference is more than 3°F, ADD refrigerant to increase the sub-cooling temperature, REMOVE refrigerant (using mandated refrigerant management practices) to decrease the sub-cooling temperature.

7. Unit Operation: Test unit operation in accordance with manufacturer's start-up procedures and repair as necessary.

4.3.1.2—Computer Room Units

Computer room units, which are designed to provide very tight control of both temperature and humidity, should be considered "mission critical" and checked monthly for proper performance, as follows:

1. Environmental Control Functions: Test by actuating each of the main functions. This is done by temporarily changing the setpoints.

2. Cooling: Test by setting the setpoint to a temperature 10°F below room temperature. A call for cooling should be seen and the equipment should begin to cool. A high temperature alarm may come on. Disregard it. Return setpoint to the desired temperature.

3. Heating: Test by setting the setpoint for 10°F above room temperature. A call for heating should be seen and the heating coils should begin to heat. Disregard the temperature alarm and return the setpoint to the desired temperature.

4. Humidification: Test by setting the humidity setpoint to a relative humidity 10% above the room humidity. For infrared humidifiers, all the infrared bulbs should come on. For steam generating humidifiers, you will immediately hear the clicks as it energizes. After a short delay, the canister will fill with water. The water will heat and steam will be produced. Return the humidity setpoint to the desired humidity.

5. Dehumidification: Test by setting the humidity setpoint to a relative humidity 10% below room relative humidity. The compressor should come on. Return humidity setpoint to the desired humidity.

6. Proportional Heating/Cooling/Dehumidification: On chilled water or glycol based units with hot water reheat, the controls are typically capable of responding to changes in room conditions. These systems utilize either a two or three-way valve activated by a proportioning motor. For cooling and dehumidification, the microprocessor will respond by positioning the valve proportionally to match the needs of the room. Full travel of the valve takes place within the range of the sensitivity setting. During dehumidification, full travel of the valve takes place within 2% RH. For hot water reheat, the microprocessor will respond by positioning the hot water valve proportionally to match the needs of the room. Full travel of the valve takes place within 1°F with each 0.1°F resulting in 10% valve travel.

Maintenance checks and service must include the following:

COMPONENT	CHECK/REMEDIAL	FREQUENCY
Filters	a. Check for unrestricted air flow.	Monthly
	b. Check filter switch	Monthly
	c. Clean filter section, replace filter.	Monthly
Fan/Coil section(s)	a. Check that fans are clean.	Monthly
	b. Lubricate bearings in accord-	Monthly

	ance with manufacturer's instructions.	
	c. Check drive belt tension and condition. Replace if required.	Monthly
	d. Inspect and tighten electrical connections.	6 Months
	e. Check coils for cleanliness and corrosion. (See Maintenance Procedure 4.3.3.3.)	Monthly
	f. Inspect drain pans. Maintain in accordance with Maintenance Procedure 4.3.3.4.	Monthly
Compressor section	a. Check for leaks.	Monthly
	b. Check oil level.	Monthly
	c. Inspect and tighten electrical connections	6 Months
Air-cooled condenser	Service in accordance with Maintenance Procedure 4.3.15.0.	Monthly
Water or glycol condenser	a. Service in accordance with Maintenance Procedure 4.3.15.1.	6 Months
	b. Inspect glycol pump for leaks.	6 Months
	c. Test pump for proper operation.	6 Months
	d. Inspect and tighten electrical connections.	6 Months
Refrigeration section	a. Check suction pressure, head pressure, and discharge pressure. Add refrigerant as needed.	Monthly
	b. Check refrigerant lines.	Monthly
	c. Check sight glass for moisture.	Monthly
	d. Check hot gas bypass valve.	Monthly
	e. Check thermostatic expansion valve.	Monthly
Reheat	a. Check reheat coil(s) operation.	Monthly
	b. Check coils for cleanliness and corrosion. (See Maintenance Procedure 4.3.3.3.)	Monthly
	c. Inspect and tighten electrical connections.	6 Months
Steam generating	a. Check canister for deposits.	Monthly

humidifier	Replace if needed.	
	b. Check condition of steam hoses.	Monthly
	c. Check water make up lines and valve for leaks.	Monthly
	d. Inspect and tighten electrical connections.	6 Months
	e. Inspect and clean out drain line.	6 Months
Infrared humidifier	a. Check drain pan for clogs.	Monthly
	b. Check humidifier lamps. Replace as necessary.	Monthly
	c. Check pan for mineral deposits. Clean as required.	Monthly
	d. Inspect and tighten electrical connections.	6 Months
Air distribution section	Clean discharge grille.	6 Months
Electrical panel	a. Check fuses.	6 Months
	b. Inspect and tighten electrical connections.	6 Months
	c. Check operation sequence.	6 Months
	d. Check contactor operation.	

4.3.2.0—Water Source Heat Pumps

Inspect and service water source heat pumps, as follows:

Airside

Inspect and maintain in accordance with Maintenance Procedure 4.3.3.2.

Waterside

1. Monthly, check and bleed air as necessary, if water system is "open loop" served from a well or water reservoir (air entrainment and binding at the water-refrigerant heat exchanger is possible). Add an inverted P-trap placed in the discharge line to keep water in the heat exchanger during off cycles and check monthly for proper operation.

2.   Clean condenser every 2 years in closed loop systems and every 6 months in open loop systems. (The cleaning cycle for open loop systems can be reduced by installing a water-to-water heat exchanger and second loop pump to effectively convert the system to a closed loop configuration.) Clean condenser as follows:

    a.   Use an inhibited hydrochloric acid solution. Cover surroundings to guard against splashing.

    b.   Gravity flow method: Valve off and drain the heat exchanger. Add cleaning solution, not any faster than vent can release gases, until coil is full. Allow solution to remain overnight (24 hours for heavy scale), then drain and flush heat exchanger with clean water.

    c.   Forced circulation method: Valve off and drain the heat exchanger. Using a portable cleaning solution tank and pump, connect to heat exchanger supply and return lines with valved connections. Fully open the vent line and fill the condenser. Then, close the vent line and start pump. Regulate flow through the condenser with the supply line valve until condenser is full, then close supply valve and stop pump. Allow solution to remain overnight (24 hours for heavy scale), then drain and flush heat exchanger with clean water.

Refrigerant side

1.   Insert a *digital thermometer* (do not use dial thermometer) under the insulation on the liquid line near the filter-dyer.

2.   Connect pressure gauge to the compressor discharge line.

3.   Operate unit for 15 minutes and read discharge (head) pressure.

4.   Using "Pressure-Temperature" chart for the refrigerant utilized, find the equivalent saturated condensing temperature.

5.   Read the liquid line temperature and subtract the saturated condensing temperature determined in Step 4. This difference is the sub-cooling temperature.

6.   Compare this temperature to the normal sub-cooling temperature listed in the manufacturer's service literature for the unit. If the

difference is more than 3°F, ADD refrigerant to increase the sub-cooling temperature, REMOVE refrigerant (using mandated refrigerant management practices) to decrease the sub-cooling temperature.

4.3.3.0—Packaged/Field-erected AHUs

Inspect packaged or field-erected air handling units, 2000 cfm or larger capacity, and perform maintenance as follows:

Weekly

1. Inspect fan motor(s) for noise, overheating, etc. Check unit for other problems including clogged condensate drains, noise, etc. and repair as needed.

Monthly

1. Inspect, replace fan belts as required. *For multi-belt sets, replace entire set, never just 1 belt.* [If belt wear is a consistent problem, consider replacing drive with the single cogged belt system.]

2. Inspect filters for loading, collapse, or other failure. Replace filters with an MERV of 10 or less in accordance with the following schedule

Filter Type/MERV		Maximum Change-out Period
1" or less throw-away	(MERV 6-8)	3 Months
2-4" pleated media	(MERV 9-10)	4 Months

Filters with an MERV of 11-16 should be replaced with the pressure drop across the filter reaches the manufacturer's recommended "dirty" level. (Thus, each filter must be have a manometer or magnahelic gauge to accurately measure pressure drop and the maintenance program must include monitoring of this condition.) However, since some filters, especially those with a prefilter, may not reach this pressure drop for years, *these filters should replaced annually even if the pressure drop has not yet reached the "dirty" criteria.*

HEPA filters (MERV 17-20) should be removed from service before the pressure drop across the filter exceeds 5" wg *when normalized to the rated flow of the filter, based on the following relationship:*

$$DP_n = DP_m (F_{rated}/F_{measured})$$

Where

DP_n = Normalized pressure drop (in. WG)

DP_m = Measured pressure drop (in. WG)

F_{rated} = Rated airflow (CFM)

$F_{measured}$ = Actual airflow (CFM)

Studies show that the tensile strength of new filter media is directly proportional to the pressure drop at which the filter shows structural failure at the pleats. Under dry conditions, the HEPA filter media will fail the required strength test after 7 to 13 years, or at an average life of 10 years. Therefore, HEPA filters operating under dry conditions should be replaced every 10 years even if the 5" wg criteria has not yet been met.

When wet, the strength of the filter media is further decreased, reducing filter life to 3 to 7 years, or an average life of 5 years. Therefore, HEPA filters operating under occasional wet conditions should be replaced every 5 years. Note that the term "wet" does not mean "soaked"—any HEPA filter that has become soaked due to water leaks, humidifier failure, or pressure-based condensation (see below) should be replaced immediately.

Occasionally, HEPA filters are installed immediately downstream of a cooling coil in an HVAC air-handler configured as a "blow through" system. In this case, the filter becomes wet almost immediately and stays wet permanently, destroying its effectiveness. The cause of this condition is condensation due to pressure reduction as the supply airstream, at or near saturation conditions leaving the cooling coil, travels through the filter. The only cures for this problem are to (a) reconfigure the AHU as a "draw through" unit or (b) add a reheat coil between the cooling coil and the filter to increase air temperature 2-3°F.

3.   Check dampers, linkage, and operators for proper operation. Dampers must be kept free of dirt or other foreign matter that may impede normal free movement. Linkage pivot points should be lubricated regularly. Axles rotate in sleeve bearings, which should not require any lubrication. It is recommended that all linkage joints be inspected periodically to insure tightness of set screws. On multizone units,

check zone dampers linkage bearings and axles to insure proper operation.

4. Check control valve(s) for proper operation.

5. Test duct smoke detector(s), freezestat, and other safety controls for proper operation.

6. Inspect fan(s) for wear and dirt. operating problems, noise, etc.
   a. Check fan bearing alignment temperature (not over 180°F).
   b. Lubricate fan bearings in accordance with Maintenance Procedure 4.3.22.0.
   c. Replace bearings if bearings or seals have failed. New bearings must have "Basic Rating Life (L_{10})" of 80,000 hours.
   d. Foundation bolts and all set screws should be inspected for tightness.
   e. Clean fan wheel as needed with a wash down with steam or water jet. Cover the bearings so water won't enter the pillow block. Dirt accumulation in the housing should be removed. Fan wheels having worn blades should be replaced.
   f. After cleaning, check fan wheel for static balance. If problems are indicated, retain fan specialist for repairs.
   g. Check V-belt drives for belt wear, alignment, and proper belt tension. Replace multiple belts when worn with complete matched set. (A better solution is replace multiple belt drives with single power belt drives.)
   h. Lubricate integral horsepower motor bearings in accordance with Maintenance Procedure 4.3.20.0.

7. If fan has inlet vanes, check main control shaft and vane mechanism. Check operating movement for range, binding, etc. Repair as required. [If inlet vanes are failing, remove inlet vanes and install variable frequency drive if motor is 20 nameplate horsepower or larger.]

8. If fan has a variable frequency drive, test drive operation in accordance with manufacturer's troubleshooting directions.

9. Coils may be cleaned in accordance with Maintenance Procedure 4.3.3.3.

10.   Inspect drain pans. Maintain in accordance with Maintenance Procedure 4.3.3.4.

4.3.3.1—Outdoor AHUs

Inspect and service outdoor air-handling units in accordance with Maintenance Procedure 4.3.3.0 and the following additional monthly requirements

1.   Check casing for corrosion or mechanical damage from limbs, etc.

2.   Check curb flashing and counter-flashing. Repair as needed

3.   Check for deterioration of flexible duct connector. Replace as needed

4.   Check for duct and/or duct insulation problems and repair as necessary.

5.   Check piping and/or pipe insulation problems and repair as necessary.

4.3.3.2—Fancoil/Blower Coil Units

Inspect fancoil units or blower-coil units with less than 2000 cfm capacity and service as follows:

Every 3 months

1.   Fan motors on these types of units are typically supplied with permanently lubricated ball bearings and will require no further lubrication. Check manufacturer's literature to determine if motor requires periodic lubrication. If so, lubricate in accordance with Maintenance Procedure 4.3.20.0.

2.   Inspect drain pans. Maintain in accordance with Maintenance Procedure 4.3.3.4.

3.   Remove filters and check inside of unit casing for dirt and lint which may accumulate around filter frame. Vacuum inside unit.

4.   Air filters: Replace quarterly with filters rated at not less than MERV 8.

5.   Check fan belt(s) tension and adjust. Replace frayed belts. Pulleys should be aligned correctly to prevent unnecessary wear on belts. *A belt dressing should never be used.* No oil or grease should come in contact with the belts as this will cause deterioration.

6.   Controls: Test controls for proper operation. Replace malfunctioning components.

Every 2 years

1.   Clean coils in accordance with Maintenance Procedure 4.3.3.3. Check coil baffles for tight fit so that air will not be allowed to bypass coils—seal as needed.

4.3.3.3—Cleaning Heating/Cooling Coils

Routinely, HVAC coils must be cleaned. There are two ways to determine if a coil requires cleaning:

1.   *Increased static pressure loss:* As a rule of thumb, at design air flow, most chilled water coils have 3/4" (dry) to 1-1/2" (wet) w.g. air pressure drop. Check the O&M manual for the design airflow and pressure drop of each coil. Measure both and evaluate actual conditions against design conditions. If the pressure drop is more than 25% greater than the equivalent design value, the coil is probably dirty.

2.   *Reduced water temperature rise/drop (low delta-T):* Test with the coil operating at "design conditions," which may require false-loading if not done during peak cooling or heating period. If supply air temperature setpoint cannot be achieved or if the water temperature rise/drop is less than design, then the coil is probably dirty.

DX Cooling and Hot Water or Steam Heating Coils

Coils can be cleaned with a vacuum cleaner, washed out with water (*no pressure washers*), blown out with low-pressure compressed air, or brushed (do *not use a wire brush*). If water is used, protect fan motors, they are drip-proof, but not waterproof. Clean coil as follows:

1.   Turn off unit power.

2.   Remove debris guards, casing, etc. to provide access to coils.

3.   Using a water hose, vacuum cleaner, or compressed air (low pressure) flush vacuum, or blow down between sections of coil to remove dirt and debris. Test cleanliness by shining a strong light through the fins from one side of the coil to the other. Repeat cleaning with a different method if coil remains dirty after first cleaning.

4.   Clean coil piping connections and surrounding casing in the same manner.

5.   *Inspect fins for aluminum oxidation and deterioration and adjust remaining unit performance life based on their condition (it is usually less expensive to replace the unit than to replace the coils).*

6.   If fins are bent or smashed, use a coil comb to straighten as much as possible.

7.   Replace casing, debris guards, etc. and restart unit.

Chilled Water Cooling Coils
    More so than DX cooling coils or heating coils, chilled water coils get dirtier and are more difficult to clean for a number of reasons.

1.   The coil is so "deep" it is impossible to clean all the way through (too many rows).

2.   The fin spacing is so fine it is impossible to clean between them (too many fins per inch).

3.   The building occupants cannot stand the smell of the chemicals often used for coil cleaning.

    Clean chilled water cooling coils as follows:
1.   Use a mild dish detergent, such as Dawn® liquid, as the cleaning agent.

2.   *Keep the supply fan in operation* to provide differential air pressure across the coil to move the flushing water and cleaning solution through the coil.

3.   *Do not use a pressure washer—it is counter-productive since it just compacts the dirt into the coil passages.*

4.   Apply the detergent liquid (about a 10% solution in water) with a pump sprayer on the upstream face of the coil, allowing it to soak in. Then, flush repeatedly with clean, cold water.

5.   Repeat this process until suds and water appear on the downstream side of the coil, with the water off the coil finally running "clear."

6.   *Inspect fins for aluminum oxidation and deterioration and adjust remaining coil performance life based on their condition.*

7.   If fins are bent or smashed, use a coil comb to straighten as much as possible.

4.3.3.4—Cooling Coil Drain Pans
Cooling coil condensate drain pans must be inspected and maintained as follows:

1.   The condensate drain trap must be cleaned prior to the beginning of the cooling season to avoid blockage by algae formation, sediment, etc. in the line.

2.   Make sure that drain pans are correctly trapped and drained. Fill the drain trap at the beginning of the cooling season.

3.   At the beginning of the cooling system, drain and clean coil drain pan. Fill the drain pan with 1-2" of water and inspect for leaks. Repair any identified leaks (if necessary, add a pan liner or mop the inside of the pan with a thick coating of bitumen to seal watertight). If pan has significantly deteriorated, plan for coil section or AHU replacement.

4.   For smaller units, where a supplemental drain pan is installed, fill the auxiliary drain pan and inspect for proper drainage, leaks, etc. Test float switch (if so equipped).

5.   During the cooling season, make sure that algaecide tablets are used liberally to keep accumulated water and drain pan clean to avoid problems with Legionella.

4.3.3.5—Fire/Smoke Dampers

Fire, smoke, and combination fire/smoke dampers installed in air-handling ductwork must be inspected, tested, and maintained in accordance with NFPA-72, NFPA 90A, NFPA-92B, and NFPA-105. The most recent editions of these standards must be checked to determine if the following minimum procedures are adequate

Fire and Combination Fire/Smoke Dampers

1.   Inspect and test each damper 1 year after installation and every 4 years thereafter (every 6 years for hospitals).

2.   For fire dampers, remove the fusible link and confirm that the damper fully closes. Ensure that no foreign penetrations (screws, flanges, etc.) penetrate the blade track. Any rusted, bent, misaligned, or damaged frame parts or blades must be repaired or replaced. If repair or replacement of individual components is not possible, replace the entire damper.

3.   If the damper operates properly, re-install the fusible link. If the link is damaged or has been painted, replace it with a new link of the same size, temperature, and load rating.

4.   Lubricate all exposed moving parts of each damper with dry lubricant in accordance with the damper manufacturer's written instructions.

Smoke and Combination Fire/Smoke Dampers

Conduct tests of smoke control systems every 6 months in accordance with NFPA-72. Each smoke and combination fire/smoke damper must be actuated and cycled as pare of the associated smoke detection system. Any damper failure must be corrected by repair or replacement.

4.3.4.0—Firetube Boilers

Steam firetube boilers must be inspected and maintained as follows:

Daily

1.   Check indicating lights and alarms.

2.   Check operating water level against reference low level for shut-off.

If operating level fluctuates erratically or changes over time, level control should be replaced.

3.　　Blowdown water column.

4.　　Perform manual boiler bottom blowdown. (Upper or skimmer blowdown should be controlled automatically by conductivity controller. Check controller for proper operation.)

5.　　Check combustion visually and check burner operation in accordance with Maintenance Procedure 4.3.4.3.

6.　　Record boiler operating conditions: operating pressure/temperature, feedwater pressure/temperature, flue gas temperature and condition (clear, hazy, smoky, dark, etc.), gas pressure, oil pressure and temperature, atomizing pressure. Compare current readings to prior readings to identify potential problems.

WHAT TO LOOK FOR	PROBABLE CAUSE
Operating pressure/ temperature	Changes are typically caused by problems with control settings, burner operation, boiler efficiency, or steam demand.
Feedwater pressure/ temperature	Reduced feedwater pressure indicates feedwater pump problem. Reduced feedwater temperature indicates problem with deaerators or feedwater heater (as applicable).
Flue gas temperature (steady load)	Increase in flue gas temperature indicates need for burner adjustment, soot on fireside surfaces, and/ or scaling on waterside surfaces.
Flue gas condition (steady load)	Gas should be clear. Cloudy, hazy, or smoky discharge indicates need for burner adjustment.
Gas pressure	Supply problems or faulty pressure regulator.
Oil pressure/temperature	Reduced oil pressure can be caused by a plugged strainer, faulty regulating valve, or air leak in suction line.

	Reduce oil temperature can indicated malfunctioning oil temperature controls or fouled heater.
Atomizing pressure	Reduced atomizing pressure indicates air supply problem (air-atomizing) or atomizing pump problem (pressure-atomizing).

7.   Check for leaks, noise, vibration, or other unusual conditions.

Weekly

1.   Check for tight closing of fuel valves.

2.   Check fuel and air linkage on burner. Tighten linkage connectors.

3.   Check operating and limit controls. If test buttons are provide, use these to test. Otherwise, simulate conditions to activate a light or alarm. Replace any control that malfunctions.

4.   Check safety and interlock controls. If test buttons are provide, use these to test. Otherwise, simulate conditions to activate a light or alarm. Replace any control that malfunctions.

5.   Check low water cutoff(s) operation and feedwater controls. Test water level control by turning off feedwater supply and let water level drop. Mark the gauge at the level when it shuts-off boiler. Use this reference for daily monthly water level checks.

6.   Check operation of all motors.

7.   Check lubricating oil levels.

8.   Check packing glands.

Monthly

1.   Inspect burner. If oil-fired, clean nozzles or cup as required.

2.   Analyze combustion and adjust air/fuel ratio as required. Use an "all-in-one" kit with chemical tubes and hand pump to sample flue gas to measure excess air, oxygen, carbon monoxide, and smoke.

3.   Inspect for flue gas leaks.

4.   Inspect and clean combustion air intake.

5.   Check fuel filters—replace as required.

Every 6 Months
1.   Remove and clean low water cut-off(s).

2.   Check oil preheater for proper operation. If fouled, remove and clean.

3.   Inspect refractory (dry back boilers). Clean cracks up to 1/8" are acceptable, larger cracks should be caulked.

4.   Replace door gaskets.

5.   Clean oil pump strainer and filter. Replace filter if required.

6.   Clean air cleaner and air/oil separator.

7.   Check fuel oil pump coupling alignment.

8.   Inspect all mercury switches and test for proper operation.

Annually
1.   Inspect and clean fireside surfaces.
    a.   Use lights to inspect the furnace and tubes for plasters or pock-marks caused by condensing flue gases. If present, implement program to keep boiler water temperature at 170°F or higher.
    b.   Look for soot. Clean if necessary.
    c.   Inspect tube sheet at tube ends for leakage, indicated by whitish streaks or deposits. If found, contact boiler service firm to have tubes rerolled.

2.   Inspect and clean waterside surfaces.
    a.   Drain boiler and remove all manhole and handhole covers.
    b.   Visually check the tubes, furnace, shell, and tube sheets for blisters, pock marks, or erosion of metal surfaces. Look for scale

formation on tubes, oxygen corrosion or pitting, and evidence of carryover water.

   c. Adjust water treatment program as indicated.

   d. Clean and de-scale if required.

3.   Inspect refractory (dry back boilers). Clean cracks up to 1/8" are acceptable; larger cracks should be caulked. Replace refractory every 5 years.

4.   Replace door gaskets.

5.   Inspect and clean breeching.

6.   Check and service feedwater pumps.

7.   Tighten all electrical terminals.

8.   Test safety relief valve by raising boiler operating pressure to the relief valve setting. Valve should open. Reduce boiler pressure. Valve should close. If valve does not open or close properly, replace valve.

9.   Disassemble, clean low water cutoff and feedwater control devices. Replace/recondition as required, reassemble, and test for proper operation.

10.   Inspect the venting system
   a. Check chimney or vent to make sure it is clean and free of cracks, open joints, or other leaks.
   b. Check that the vent connector is inserted into the chimney, but not extending beyond the inside edge of the chimney.
   c. Check and clean combustion air intake.

11.   Inspect the boiler room
   a. The boiler room must be clean from combustibles or flammable liquids or vapors.
   b. Remove all stored materials from boiler room,

Every 5 years

1.   Replace refractory and door gaskets on dry back boilers.

4.3.4.1—Cast Iron Boilers

Cast iron boilers must be inspected and maintained as follows:

Seasonal start-up

1. Inspect the venting system
    a. Check chimney or vent to make sure it is clean and free of cracks, open joints, or other leaks.
    b. Check that the vent connector is inserted into the chimney, but not extending beyond the inside edge of the chimney.
    c. Check and clean combustion air intake.

2. Inspect the boiler room
    a. The boiler room must be clean from combustibles or flammable liquids or vapors.
    b. Remove all stored materials from boiler room,

3. Start boiler
    a. For steam boilers, disassemble and clean low water cutoff and feedwater control devices. Replace/recondition as required, reassemble, and test for proper operation.
    b. Start-up boiler/burner in accordance with manufacturer's written instructions.
    c. Test for correct operation of ignition and flame proving controls.
    d. Test for correct operation of low water cutoffs (steam boiler)
    e. Test for correct operation of safety relief valve(s) in accordance with Maintenance Procedure 4.3.18.4.2 (steam) or 4.3.18.4.5 (water).
    f. Visually inspect burner in accordance with Maintenance Procedure 4.3.4.3 (pressure burner) or 4.3.4.4 (atmospheric burner).

Monthly

1. Inspect the boiler room
    a. The boiler room must be clean from combustibles or flammable liquids or vapors.
    b. Remove all stored materials from boiler room,

2. Inspect the venting system
    a. Check chimney or vent to make sure it is clean and free of cracks, open joints, or other leaks.

b. Check and clean combustion air intake.

3.   Inspect boiler, burner, and controls for proper operation.

4.   Check the boiler and piping for any leaks (water or steam) and repair as necessary.

Seasonal shutdown

1.   Turn off the burner.

2.   Open the main power disconnect switch to the boiler/burner. Close fuel supply valve(s).

3.   Clean all metal surfaces of the firebox, gas passages, and venting (breechings, vent pipe, etc.) with a wire brush and vacuum.

4.   Check condition of boiler surface. Clean any rust, deposits, etc. and paint with high temperature protective coating to reduce corrosion.

5.   Examine sections for proper expansion allowance between sections. (For push-nipple type boilers, make sure tie rods are removed or are loose enough to allow expansion without cracking sections.)

6.   Check for evidence of leaks between sections. Disassemble sections, clean off corrosion, and reassemble, repacking the joints between sections with insulating material. (For push-nipple type boilers, be sure to allow for thermal expansion between sections— see manufacturer's assembly instructions.)

7.   Examine condition of boiler water. If the water has considerable foreign matter in it, drain and flush the boiler by removing the front and rear sections' bottom plugs.

8.   Fill the boiler with clean water. Steam boilers should be filled to the top of the steam outlet connection.

4.3.4.2—Electric Hot Water Boilers

Electric hot water boilers must be inspected and maintained as follows:

Seasonal start-up
1. Inspect the boiler room
   a. The boiler room must be clean from combustibles or flammable liquids or vapors.
   b. Remove all stored materials from boiler room,

2. Start boiler
   a. Clean contactors and tighten wiring connections. Check contactor coils. Test fuses and replace as required.
   b. Test for correct operation of safety relief valve(s) in accordance with Maintenance Procedure 4.3.18.4.5.
   c. Close main disconnect switch and energize boiler in accordance with the manufacturer's written instructions.

Monthly
1. Inspect the boiler room
   a. The boiler room must be clean from combustibles or flammable liquids or vapors.
   b. Remove all stored materials from boiler room,

3. Inspect boiler and test controls for proper operation.

4. Check the boiler and piping for any leaks and repair as necessary.

Seasonal Shutdown
1. Open main disconnect, lock and tag.

2. Take boiler water sample for analysis. If boiler water has high TDS, drain and refill system. Add water treatment chemicals for a closed system.

4.3.4.3—Pressure Type Burners

Inspect, test, and maintain pressure type burners (gas-, oil-, or dual-fuel) as follows:

WHAT TO LOOK FOR	FREQUENCY	REMEDIAL ACTIONS
Gauges, monitors and indicators	Daily	Make visual inspection and record readings
Instruments and equipment settings	Daily	Make visual check against manufacturer's specifications
Atomizing air	Weekly	For air atomizing burners, check filter and drain moisture from traps. For pressure atomizing burners, make sure oil pressure is correct.
Firing rate control	Monthly	Test firing efficiency with combustion test. Tune burner as required for maximum efficiency on both high and low fire.
Flue, vent or stack, outlet damper (if multiple boilers)	Monthly	Make visual inspection, check damper linkage and operation
Combustion air	Monthly	Check that intake louvers are clear and any dampers work correctly.
Ignition system	Weekly Monthly	Make visual inspection. Check flame proof signal strength.
Fuel valves, pilot and main	Weekly	Open limit switch, make sound and visual check and check valve position indicators.
Fuel valves, pilot and main	Annually	Perform leakage tests.
Combustion and safety controls, flame failure	Weekly	Close manual fuel supply for pilot and main fuel cock/valves, check safety shutdown timing

Combustion and safety controls, flame signal strength	Weekly	If a flame signal meter is installed, read and log for both pilot and main flames; check for high, very low, or fluctuating readings.
Combustion and safety controls, pilot turndown tests	Annually	Required after any adjustments to flame scanner mount or pilot burner; verify annually
High limit safety control	Annually	Test for proper function
Operating control	Annually	Test for proper function
Low draft, fan, air pressure and damper	Monthly	Test for proper function
High and low gas pressure interlocks	Monthly	Test for proper function
Low oil pressure interlocks	Monthly	Test for proper function
Fuel valve interlock switch	Annually	Test for proper function
Purge switch	Annually	Test for proper function
Low fire start interlock	Annually	Test for proper function
Dual fuel changeover controls	6 Months	Test for proper function
Oil filter	Monthly	Remove and clean; replace if required
Blower, motor and wheel	Annually	Inspect and clean; lubricate in accordance with Maintenance Procedures 4.3.20.0 and 4.3.22.0.
Gas pilot assembly	Annually	Inspect and clean

4.3.4.4—Atmospheric Gas-fired Burners

Monthly, inspect and service atmospheric gas-fired burners.

1. Inspect gas flame. If it is not blue (i.e., orange or yellow), burner adjustment is required.

2. Shut off fuel supply, open access panel(s), and clean burner ports using stiff wire or needle.

3. Inspect pilot assembly (if applicable)
    a. Check wiring connections for tightness.
    b. Check electrode, flame sensor, and pilot flame. If any component is loose or if ceramic insulators are cracked or broken, replace assembly.

4. Inspect and test spark ignition system (if applicable).
    a. Test "Flame Failure" control. Close the downstream gas shutoff valve and confirm that safety shutdown occurs. If not, replace control.
    b. Test "Flame Signal" strength in accordance with manufacturer's written instructions.

5. Inspect the flue assembly for leaks. Replace any corroded vent piping, seal joints, adjust hangers, etc.

6. Every 6 months, test the pressure/temperature relief valve by lifting lever or steam. If valve malfunctions or shows significant corrosion or deposits, replace.

4.3.5.0—Gas/Oil-fired Furnaces

Annually, at the beginning of each heating season, inspect and service gas- or oil-fired furnaces.

1. Inspect and maintain airside of furnace in accordance with Maintenance Procedure 4.3.3.2.

2. Inspect and maintain burner in accordance with Maintenance Procedure 4.3.4.4.

3. Inspect the venting system

a. Check chimney or vent to make sure it is clean and free of cracks, open joints, or other leaks.

b. Check that the vent connector is inserted into the chimney, but not extending beyond the inside edge of the chimney.

c. Check and clean combustion air intake.

4.3.6.0—Gas/Oil-fired Unit Heaters
Annually, inspect and service gas- or oil-fired unit heaters.

1. Fan motors on these types of units are typically supplied with permanently lubricated ball bearings and will require no further lubrication. Check manufacturer's literature to determine if motor requires periodic lubrication. If so, lubricate in accordance with Maintenance Procedure 4.3.20.0.

2. Test controls for proper operation. Replace malfunctioning components.

3. Clean coils with a stiff brush and vacuum cleaner or compressed air. Check coil baffles for tight fit so that air will not be allowed to bypass coils—seal as needed.

4. Inspect and maintain burner in accordance with Maintenance Procedure 4.3.4.4

5. Inspect the venting system
    a. Check chimney or vent to make sure it is clean and free of cracks, open joints, or other leaks.
    b. Check that the vent connector is inserted into the chimney, but not extending beyond the inside edge of the chimney.
    c. Check and clean combustion air intake.

4.3.6.1—Hot Water/Steam Unit Heaters
Annually, inspect and service hot water or steam unit heaters.

1. Fan motors on these types of units are typically supplied with permanently lubricated ball bearings and will require no further lubri-

cation. Check manufacturer's literature to determine if motor requires periodic lubrication. If so, lubricate in accordance with Maintenance Procedure 4.3.20.0.

2. Test controls for proper operation. Replace malfunctioning components.

3. Clean coils with a stiff brush and vacuum cleaner or compressed air. Check coil baffles for tight fit so that air will not be allowed to bypass coils—seal as needed.

4. For steam heaters
    a. Valve off strainer and remove screen. Clean and reassemble.
    b. Verify that steam trap is working properly. A blocked steam trap will be cold. A trap that has failed open will be hot and noisy, and the discharge pipe will be hot for up to 30 feet downstream of the trap. If trap has failed, replace it.

5. Inspect control valve for (a) tight shutoff against and (b) proper function. Check for steam or water leaks. Check pneumatic actuators for air leaks.

4.3.6.2—Electric Unit Heaters
Annually, inspect and service electric unit heaters.

1. Fan motors on these types of units are typically supplied with permanently lubricated ball bearings and will require no further lubrication. Check manufacturer's literature to determine if motor requires periodic lubrication. If so, lubricate in accordance with Maintenance Procedure 4.3.20.0.

2. Inspect all electrical components
    a. Make sure connections are tight.
    b. Check fuses and replace any that are blown.
    c. Check contactors for proper function, burned contacts, etc. Replace as required.
    d. Check for required clearance in front of electrical service panel.

3. Inspect heater elements. If elements are corroded (due to salt or

moisture) or broken or ceramic bushings are cracked or broken, re-place coil. If elements are dirty, clean with low pressure compressed air and a brush duster, being careful not to damage coils.

4.   Check heater operation
   a.   Check installation instructions and wiring diagrams to make sure heater is wired and installed properly.
   b.   If automatic high-limit switch or airflow switch has tripped, temperature may be too high because airflow is insufficient or uneven. If energized for areas that glow "red," low or uneven airflow exists. If airflow switch "chatters," low airflow exists.

4.3.7.0—Gas-fired Radiant Heaters

Annually, before the heating season, service gas-fired radiant heaters as follows:

1.   Disconnect electric power and close gas supply valve.

2.   With 15 psig (maximum) compressed air, blow off dust and dirt on burner assembly, tubes, and reflectors. Use dry cloth if required, but do not use chemical cleaners or water. Do not insert air hose into inlet of each venturi tube.

3.   Remove burner orifices from header manifold, clean with drill size(s) specified by manufacturer, and re-install.

4.   Remove pilot orifice, clean with drill size specified by manufacturer, and re-install.

5.   Check wiring for loose connections and tighten as required. Replace any wiring with damaged or deteriorated insulation.

6.   Inspect the venting system
   a.   Check chimney or vent to make sure it is clean and free of cracks, open joints, or other leaks.
   b.   Check that the vent connector is inserted into the chimney, but not extending beyond the inside edge of the chimney.

7.   Re-connect electric power and open gas supply valve.

8.   Test controls for proper operation. Replace malfunctioning components.

4.3.8.0—Terminal Units

Annually (every 6 months for units serving critical areas such as hospital patient and treatment areas, labs, etc.), check terminal units for proper operation in accordance with the following:

1.   Open terminal unit and check damper for binding, sticking, etc. Adjust as needed.

2.   Make sure damper motor is working—check through full control range. Replace if needed.

3.   Check and adjust damper linkage.

4.   Clean velocity sensor array.

5.   Vacuum inside of terminal unit.

6.   Clean reheat coil (if present) with a stiff brush and vacuum cleaner or compressed air. Check coil baffles for tight fit so that air will not be allowed to by-pass coil—seal as needed.

7.   Test controls for proper operation. Calibrate sensors, tune control loop, and/or replace malfunctioning components as needed.

4.3.8.1—Fan-powered Terminal Units

Annually, (every 6 months for units serving critical areas such as hospital patient and treatment areas, laboratories, etc.), check fan-powered terminal units for proper operation in accordance with the following:

1.   Open terminal unit and check damper for binding, sticking, etc. Adjust as needed.

2.   Make sure damper motor is working—check through full control range. Replace if needed.

3.   Check and adjust damper linkage.

4.   Clean velocity sensor array.

5.   Vacuum inside of terminal unit.

6.   Lubrication: Fan motors on these types of units are typically supplied with permanently lubricated ball bearings and will require no further lubrication. If motor requires periodic lubrication, do so in accordance with Maintenance Procedure 4.3.20.0.

7.   Clean reheat coil (if present) with a stiff brush and vacuum cleaner or compressed air. Check coil baffles for tight fit so that air will not be allowed to by-pass coil—seal as needed.

8.   Check fan belt(s) tension and adjust. Replace frayed belts. Pulleys should be aligned correctly to prevent unnecessary wear on belts. A belt dressing should never be used. No oil or grease should come in contact with the belts as this will cause deterioration.

9.   Test controls for proper operation. Calibrate sensors, tune control loop, and/or replace malfunctioning components as needed.

4.3.9.0—Return & Exhaust Air Duct Systems
Inspect return and exhaust air duct systems once a year (install additional access doors or panels as needed for inspection and duct cleaning).

1.   Visually inspect duct interior areas for potential moisture accumulation, including first 20 feet of supply duct and outdoor air and mixed air plenums, for biological growth. If present, clean and disinfect.

2.   Inspect exposed ductwork for insulation and vapor barrier integrity and repair as required. Check for sagging or water-logged insulation and replace as required.

3.   Evaluate the need for duct system cleaning. Systems may need cleaning every 5-10 years.

4.3.9.1—Registers, Grilles, Diffusers
Every 6 months, check registers, grilles, and diffusers for air pattern changes, rattles or noise, etc. that may indicate that a damper or

flow control grid has broken. Clean as required. (Note: Ceiling smudging is an indication that the air distribution device is working correctly, but the building is dirty. Address required cleanliness with housekeeping.)

4.3.9.2—Humidifiers

Inspect and maintain humidifiers as follows:

Steam injection humidifiers
Every 6 months
1. Valve off strainer and remove screen. Clean and reassemble.

2. Verify that steam trap is working properly. A blocked steam trap will be cold. A trap that has failed open will be hot and noisy, and the discharge pipe will be hot for up to 30 feet downstream of the trap. If trap has failed, replace it.

3. Inspect control valve for (a) tight shutoff against steam pressure and (b) proper function. Check for leaks. Check pneumatic actuators for air leaks.

Annually
Inspect silencer for cleanliness. Clean or replace as required.

Every 2 years
Replace seal rings and O-rings.

Steam-to-steam humidifiers
Monthly
1. Check humidifier tank for scale build-up. Remove scale before it reaches the underside of the heat exchanger(s).

2. Remove sensing probe and clean scale (which should easily flake off).

3. Clean skimmer tube. If mineral deposits restrict flow, replace.

4. Inspect gaskets and replace as needed.

Every 3 months

1. Disassemble humidifier and clean. If tank has Teflon liner, be very careful not to scratch.

2. Adjust the surface skim or bleed-off in accordance with manufacturer's written instructions.

3. After reassembly, check for leaks.

Every 6 months

1. Valve off steam supply strainer and remove screen. Clean and reassemble.

2. Verify that steam trap is working properly. A blocked steam trap will be cold. A trap that has failed open will be hot and noisy, and the discharge pipe will be hot for up to 30 feet downstream of the trap. If trap has failed, replace it.

3. Inspect control valve for (a) tight shutoff against steam pressure and (b) proper function. Check for leaks. Check pneumatic actuators for air leaks.

Electric steam humidifiers
Monthly

1. Check humidifier tank for scale build-up. Remove scale before it reaches the underside of the heat exchanger(s).

2. Remove sensing probe and clean scale (which should easily flake off).

3. Clean skimmer port. If mineral deposits restrict flow, replace piping. Check for proper skimmer flow by visual inspection.

Every 3 months

1. Disassemble humidifier and clean. If tank has Teflon liner, be very careful not to scratch.

2. Adjust the surface skim or bleed-off in accordance with manufacturer's written instructions.

3. After reassembly, check for leaks.

4.3.10.0—Fans (2000 cfm)

Inspect and maintain fans rated at more than 2000 cfm capacity

Monthly

1. Inspect fan(s) for wear and dirt. operating problems, noise, etc.
   a. Check fan bearing alignment temperature (not over 180°F).
   b. Lubricate fan bearings in accordance with Maintenance Procedure 4.3.22.0.
   c. Replace bearings if bearings or seals have failed. New bearings must have "Basic Rating Life (L_{10})" of 80,000 hours.
   d. Foundation or hanger bolts, rods, and set screws should be inspected for tightness.
   e. Clean fan wheel as needed with a wash down with steam or water jet. Cover the bearings so water won't enter the pillow block. Dirt accumulation in the housing should be removed. Fan wheels having worn blades should be replaced.
   f. After cleaning, check fan wheel for static balance. If problems are indicated, retain fan specialist for repairs.
   g. Check V-belt drives for belt wear, alignment, and proper belt tension. Replace multiple belts when worn with complete matched set. [If belt wear is a consistent problem, replace multiple belt drive with single power belt drive.]
   h. Lubricate integral horsepower motor bearings in accordance with Maintenance Procedure 4.3.20.0.

2. If fan has inlet vanes, check main control shaft and vane mechanism. Check operating movement for range, binding, etc. Repair as required. [If inlet vanes are failing, remove inlet vanes and install variable frequency drive if motor is 20 nameplate horsepower or larger.]

3. If fan has a variable frequency drive, test drive operation in accordance with manufacturer's troubleshooting directions.

4.3.10.1—Exhaust Fans

Inspect and maintain exhaust fans as follows:

Weekly

1. Check that fans are operational. Inspect fan(s) for broken belts, wear and dirt, noise, etc.

Monthly
1. Check fan bearing alignment temperature (not over 180°F).

2. Lubricate integral horsepower motor bearings in accordance with Maintenance Procedure 4.3.20.0.

3. Foundation or hanger bolts, rods, and set screws should be inspected for tightness.

4. Clean fan wheel as needed with a wash down with steam or water jet. Cover the bearings so water won't enter the pillow block. Dirt accumulation in the housing should be removed. Fan wheels having worn blades should be replaced.

5. Check V-belt drives for belt wear, alignment, and proper belt tension. Replace multiple belts when worn with complete matched set.

4.3.11.0—Electric Duct Heaters
Annually, inspect and service electric duct heaters.
1. Inspect all electrical components
    a. Make sure connections are tight.
    b. Check fuses and replace any that are blown.
    c. Check contactors for proper function, burned contacts, etc. Replace as required.
    d. Check for required clearance in front of electrical service panel.

2. Inspect heater elements [add duct access door(s) as necessary]. If elements are corroded (due to salt or moisture) or broken or ceramic bushings are cracked or broken, replace coil. If elements are dirty, clean with low pressure compressed air and a brush duster, being careful not to damage coils.

3. Check heater operation
    a. Check installation instructions and wiring diagrams to make sure heater is wired and installed properly.
    b. If automatic high-limit switch or airflow switch has tripped, temperature may be too high because airflow is insufficient or uneven. If energized for areas that glow "red," low or uneven airflow exists. If airflow switch "chatters," low airflow exists.

c. Check transformers and control voltage.

d. If an air pressure switch is used, check that sensing tubes are curved toward the airflow (facing upstream).

e. Check that duct liner is not touching coil or interfering with any safety device.

f. Utilized following troubleshooting guide to diagnose heater operating problems

WHAT TO LOOK FOR	PROBABLE CAUSE	REMEDIAL ACTIONS
Heater does not operate	No power	Check disconnect or panel breaker
		Check power fuse(s), replace as necessary
	No control voltage	Check control voltage source (if independent)
		Check transformer and transformer fusing, replace as necessary
	Open limit switch (primary or secondary)	Reset
		Check for continuity across limit switch to determine if open, replace as necessary
	Low static pressure	Increase fan static pressure setting
		For VAV, make sure static pressure at fan minimum flow rate is adequate, modify fan control as necessary
	Damaged heating element	Check for open or damaged element(s), replace as necessary
High or low temperature rise	Incorrect airflow	See below
	Thermostat or controller failure	Check wiring
		Check thermostat or controller operation, replace as necessary
	Staging problem(s)	Check heat staging control by controller, replace as necessary
		Check contactors for open

		coil, replace as necessary
		Check for damage elements (see above)
Short Cycling	Incorrect airflow	See below
Incorrect airflow	Fan or duct problems	Check for even airflow across the face of each heater element
		Check for minimum airflow of 70 cfm per kW of coil rating
		Check for blocked duct (fire damper, insulation, etc.) or fan problems
		Check for dirty filters
Heater with SCR controller does not function	Controller failure	Test controller, using manufacturer's troubleshooting guide, replace as necessary

4.3.12.0—Steam/Hot-water Heat Exchangers

Inspect and service steam-to-hot-water or hot-water-to-hot-water heat exchangers annually as follows:

1. Do not open exchanger for inspection until all pressure is off and unit is completely drained.

2. Flush shell and tube bundle with clean water. If further cleaning is necessary, remove head and bundle to clean inside of shell and outside of tubes. Replace gaskets if necessary.

    If unit is installed in a hard water area, inside of tubing can be cleaned as follows:
    a. Break water connections and plug bottom opening.
    b. Fill the tubes with a solution of 1 part Muriatic acid to 10 parts water and allow to stand for 2 hours, *but no longer*. [A longer period may cause damage to the copper tubing.]
    c. Drain and flush thoroughly with clean water.
    d. Reassemble.

3. Inspect and test control and relief valves to ensure their proper operation. *If relief valve fails to function properly, replace it; do not attempt to repair.*

4.   On steam heat exchanger, check and clean steam trap and strainer on condensate outlet line.

4.3.12.1—Plate-and-frame Heat Exchangers
Inspect and service plate-and-frame heat exchangers as follows:

1.   Monthly, inspect heat exchanger for the following conditions.
   a.   Leaks that cannot be stopped by tightening bolts in accordance with manufacturer's written instructions.
   b.   Increase in pressure drop across the heat exchanger.
   c.   Reduction in heat exchange capacity indicated by reduction in temperature drop or rise.

2.   Clean heat exchanger if any of the above conditions are found.
   a.   Stop flows through heat exchanger and isolate it by closing service valves.
   b.   Allow heat exchanger to sit for 24 hours to come to ambient temperature. Do not rapidly cool or heat the exchanger as this will cause gasket leaks.
   c.   Disassemble the heat exchanger in accordance with manufacturer's written instructions.
   d.   Clean plates using pressure washing and, if necessary, non-abrasive scrubbing.
   e.   Reassemble the heat exchanger with new ethylene propylene rubber (EPDM) gaskets in accordance with manufacturer's written instructions.
   f.   Bring the heat exchanger to operating temperature slowly over a 24-hour period.

3.   Annually, blow down the strainers on both heat exchanger piping loops.

4.3.13.0—Air-cooled Electric-drive Water Chillers
Maintenance requirements for air-cooled electric-drive (rotary or reciprocating compressor) water chillers are as follows:

Daily monitoring/visual inspection
The majority of chiller operating problems and maintenance needs are discovered by visual inspection and frequent monitoring of equipment operating parameters.

Monthly, quarterly, and annual preventative maintenance
   1. Clean evaporator every 1-3 years (annually for chillers serving air washers or other "open" cooling loads). [To facilitate this cleaning, piping connections at the chiller should be flanged or use grooved piping connections and the insulation should be designed for removal and replacement. The piping must be configured so as not to interfere with the tube cleaning space. If the original installation does not meet these requirements, then piping changes must be made.]

   2. Quarterly, calibrate pressure, temperature, and flow controls.

   3. Annually, inspect starter wiring connections, contacts, and action. Tighten and adjust as require. Perform thermographic survey every 5 years.

   4. Annually, test operation of safety interlocks devices, such as flow switches, pump starter auxiliary contracts, phase-loss protection, etc. Repair or replace as required.

   5. Annually, perform dielectric motor testing to identify failures in motor winding insulation. For large chillers (100 tons or larger), additional annual motor tests are required to test for imbalance of electrical resistance among windings, imbalance of total inductance with phase inductances, power factor, capacitance imbalance, and running amperage vs. nameplate amperage.

   6. Annually, check tightness of hot gas valve (as applicable). If the valve is does provide tight shutoff, replace it.

   7. Annually, change lubricant (oil) filter and drier.

   8. Laboratory analysis of lubricant should be performed annually during the first ten years of chiller life and every six months thereafter. [This oil analysis will define the moisture content (not to exceed 50 ppm), oil acidity (maximum 1 ppm) that may indicate oil oxidation and/or refrigerant degradation due to high temperatures, and metals or metal oxides that indicate chiller component wear and/or moisture in the oil.]

9. Valve and bearing inspection in accordance with manufacturer's recommendation.

10. Relief valves (both refrigerant and water) should be checked annually. Disconnect the vent piping at the valve outlet and visually inspect the valve body and mechanism for corrosion, dirt, or leakage. If there are problems, *replace the valve, do not attempt to clean or repair it.*

11. Every 6 months, inspect and clean air-cooled condensers in accordance with Maintenance Procedure 4.3.15.0

12. Perform chemical testing of system water at least quarterly. Treat as needed to ensure proper water chemistry.

Extended (long term) maintenance checks
1. Every 5 years (or more frequently as the chiller ages), perform eddy current (electromagnetic) testing of heat exchanger tubes. [This testing will typically detect tube pits, cracks, and tube wear (thinning).] If only 1-5 tubes is found defective, plug tubes. If more than 5 are found defective, replace tubes. [Any tube that is replaced as result of this testing should be examined and cross-sections cut so the cause of the defects can be evaluated.]

2. Other components must be serviced, inspected, and/or replaced at the intervals recommended by the chiller manufacturer.

4.3.13.1—Water-cooled Electric-drive Water Chillers
Maintenance requirements for water-cooled electric-drive rotary compressor water chillers are as follows:

Daily monitoring/visual inspection
The majority of chiller operating problems and maintenance needs are discovered by visual inspection and frequent monitoring of equipment operating parameters.

Monthly, Quarterly, and Annual Preventative maintenance
1. Clean evaporator every 1-3 years (annually for chillers serving air washers or other "open" cooling loads). [To facilitate this

cleaning, piping connections at the chiller should be flanged or use grooved piping connections and the insulation should be designed for removal and replacement. The piping must be configured so as not to interfere with the tube cleaning space. If the original installation does not meet these requirements, then piping changes must be made.]

2. Clean condenser every year. [To facilitate this cleaning, piping connections at the chiller should be flanged or use grooved piping connections and the insulation should be designed for removal and replacement. The piping must be configured so as not to interfere with the tube cleaning space. If the original installation does not meet these requiremnts, then piping changes must be made.]

3. Quarterly, calibrate pressure, temperature, and flow controls.

4. Annually, inspect starter wiring connections, contacts, and action. Tighten and adjust as required. Perform thermographic survey every 5 years.

5. Annually, test operation of safety interlocks devices, such as flow switches, pump starter auxiliary contracts, phase-loss protection, etc. Repair or replace as required.

6. Annually, perform dielectric motor testing to identify failures in motor winding insulation. For large chillers (100 tons or larger), additional annual motor tests are required to test for imbalance of electrical resistance among windings, imbalance of total inductance with phase inductances, power factor, capacitance imbalance, and running amperage vs. nameplate amperage.

7. Annually, check tightness of hot gas valve (as applicable). If the valve does not provide tight shutoff, replace it.

8. Annually, replace lubricant (oil) filter and drier.

9. Laboratory analysis of lubricant should be performed annually during the first ten years of chiller life and every six months

thereafter. Analysis must address oil moisture content, acidity, and chiller wear, as follows:

a. Maximum moisture content of 50 ppm. Higher levels may indicate air leaks in low pressure chillers or heat exchanger tube leaks.

b. Maximum acidity of 1 ppm is normal for new or re-claimed refrigerant. Higher levels may be caused by oxidation of oil during aging and/or degradation of refrigerant. If higher levels are found, contact the chiller manufacturer.

c. Trace amounts of metals could be metal oxides formed from moisture in the oil or may indicate excess wear conditions. Due to variations in metals used by different chiller manufacturers, there are no standard limits. However, *if metal content increases from year to year, one or more of the following excess wear conditions by be responsible:*

Aluminum:	Impellor or bearing wear.
Chromium:	Wear on re-chromed shafts.
Copper:	Corrosion.
Iron:	Corrosion and/or gear wear.
Tin:	Bearing wear.
Silicon:	Leakage of silica gel from dehydrator cartridge or dirt in system.
Zinc:	Zinc left over from the manufacturing process or lose of galvanizing from some parts.

10. Valve and bearing inspection in accordance with manufacturer's recommendation.

11. Relief valves (both refrigerant and water) should be checked annually. Disconnect the vent piping at the valve outlet and visually inspect the valve body and mechanism for corrosion, dirt, or leakage. If there are problems, *replace the valve, do not attempt to clean or repair it.*

12. Annually, inspect gearbox for wear and repair or replace it as needed.

13. Perform chemical testing of system water at least quarterly. Treat as needed to ensure proper water chemistry.

Extended (long term) maintenance checks

1. Every 5 years (or more frequently as the chiller ages), perform eddy current (electromagnetic) testing of heat exchanger tubes. [This testing will typically detect tube pits, cracks, and tube wear (thinning).] If only a limited number of tubes are found defective, plug tubes. If more than few are found defective, replace tubes. [Any tube that is replaced as result of this testing should be examined and cross-sections cut so the cause of the defects can be evaluated.]

2. Every 3-5 years, perform vibration test and analysis to evaluate motor and rotor balance, bearing and gear alignment, and bearing and gear wear. Use accelerometer type sensors for gears and bearings (high frequency) and piezoelectric velocity sensors for compressor motors, rotors, and bearings (low velocity).

3. Every 5 years, perform an acoustic emission test to identify potential stress cracks in pressure vessels, tubes, and tube sheets.

4. Other components must be serviced, inspected, and/or replaced at the intervals recommended by the chiller manufacturer.

4.3.13.2—Absorption Water Chillers

Absorption 2-stage direct-fired water chillers have five significant maintenance elements:

Daily monitoring/visual inspection

The majority of chiller operating problems and maintenance needs are discovered by visual inspection and frequent monitoring of equipment operating parameters.

Mechanical components

1. Preferably daily, but not exceeding weekly, the operation of the purge unit must be checked for both proper operation and *excess* operation, indicating an air leak that can produce serious problems due to corrosion, contamination of the absorbent solution, and reduction in efficiency and capacity. Test for non-condensables (hydrogen).

2. Annually, refrigerant and solution pumps and the purge unit must be inspected and tested to insure continued operation. Every 3

years, the pump seals and bearings will require inspection and, as indicated, replacement.

Heat transfer components

1.   Annually, the condenser and absorber heat exchanger tubes must be cleaned. The evaporator tubes should be cleaned every 5 years. To facilitate this cleaning, piping connections at the chiller should be flanged or use grooved piping connections and the insulation should be designed for removal and replacement. The piping must be configured so as not to interfere with the tube cleaning space. If the original installation does not meet these requirements, then piping changes must be made.

2.   Every 6 months (more frequently if runtime exceeds 4000 hours per year), the lithium bromide solution must be analyzed for contamination, solids, pH, corrosion-inhibitor level, and performance additive (typically octyl alcohol). Adjust solution as necessary to meet manufacturer's requirements. [Metals and metal oxides in the solution are typical indications that internal corrosion is occurring due to air leaks.]

3.   Annually, perform leak test, using the pressure method.

4.   Annually, perform eddy current testing of the high stage generator tubes.

5.   Every 3- 5 years, perform eddy current testing of the absorber, condenser, low stage generator, and evaporator tubes.

6.   Every 3 years, service valves containing rubber diaphragms. These diaphragms should be replaced (which requires that the lithium bromide/water solution be removed from the chiller).

Controls

Annually, test the controls for proper operation at the beginning of the cooling season. Clean and tighten all connections, including field sensor connections. Vacuum control cabinets to remove dirt and dust. With microprocessor controls, have the factory service technicians test calibration and operation and ensure that the latest version of operating software is loaded.

Burner

Inspect, test, and maintain in accordance with Maintenance Procedure 4.3.4.2.

4.3.14.0—Cooling Towers

Cooling towers require the following preventative maintenance measures:

Start-up

When the tower is to be started after seasonal shutdown, it must be thoroughly inspected and repaired, as follows:

1. Check drift eliminators for proper position, being clean, etc.

2. Check fans, bearings, motors, and drives for proper lubrication.

3. Rotate fan shaft(s) by hand to make sure they turn freely.

4. Check fan motors for proper rotation and adjust belt tension for belt drives.

5. Fill basin with fresh water and check operation of level controller.

6. Start condenser pump and check wet deck for proper distribution.

7. Check fill for fouling and/or clogging and clean or replace if necessary.

8. Check access door gaskets and replace as necessary.

9. Thoroughly inspect all metal surfaces for corrosion, scale or fouling, or sludge. Clean as required. Any damaged metal should be cleaned down to bare metal and refinished with a cold zinc coating.

10. Operate tower and look for and repair any water or air leaks from the basin, casing, or piping.

Scheduled

1. *Weekly*
    a. Clean basin strainer

    b. Check blow down valve and make-up water valves to make sure they're working properly

    c. Test water and adjust chemical treatment as necessary

    d. Check/fill gear drive oil reservoir (on gear-drive towers)

2. *Monthly*
    a. Clean and flush basin. (This may be required more often for towers located adjacent to highways, industrial sites, etc. with high particulate emissions or in hot, humid climates with high bio-fouling potential.)

    b. Check operating level in basin and adjust as necessary.

    c. Check water distribution system and sprays.

    d. Check drift eliminators for proper position.

    e. Check belts or gearbox and adjust as necessary.

    f. Check fans, inlet screens, and louvers for dirt and debris. Clean as necessary.

    g. Check keys and tightness of set screws.

3. *Routinely*
    a. Lubricate TEFC motor in accordance with Maintenance Procedure 4.3.20.0. TEAO motors are sealed and do not require lubrication.

    b. Lubricate fan shaft bearings in accordance with Maintenance Procedure 4.3.22.0.

    c. Check gear drive in accordance with manufacturer's instructions. Change gear drive oil every 2500 hours of operation.

    d. Every 3 months, blowdown (Y-type) or clean (basket type) condenser water pump strainer.

4. *Yearly*
    a. Clean and touch-up paint or other protective finish as necessary (including tower support steel).

    b. Dismantle and clean condenser water pump strainer.

Water Treatment:

1. Provide automatic condenser water treatment in accordance with Maintenance Procedure 4.3.21.1.

2. Require and evaluate regular, frequent reports by the water treatment contractor or staff, first to ensure that regular water treatment is being done and, second, to "track" the various treatment parameters pH, total dissolved solids (TDS), chemical types and quantities used, etc.

3. At least twice each year, send a condenser water sample to an independent laboratory for analysis and compare the results with the most recent monthly reports provided by the water service contractor or staff.

4. During shutdown periods, inspect the tower and as much piping as possible for scaling or fouling that is being inadequately addressed by the water treatment program. Adjust program accordingly.

5. Log tower and chiller performance information at least once per day or once per shift and record the relationships between load, temperature difference, and power input. If the relationships between these values change appreciably, this could indicate chiller fouling, tower fouling, or other performance problems that require further investigation and service.

4.3.15.0—Air-cooled Condenser

Every 6 months, inspect air-cooled condenser and service as follows:

1. Check unit casing and clear any leaf litter or organic matter in contact with the casing. Remove leaves, sticks, etc on or in the unit casing. Check for condition of paint, metal, etc. and repair as necessary.

2. Check outdoor fans for proper rotation and that fans do not run backwards when off. Clean and, if needed, balance fan blades.

3. Lubrication: Typically, fan motors have sealed bearings and require no lubrication. If motor does require lubrication, do so in accordance with Maintenance Procedure 4.3.20.0.

4. Inspect condenser coils and clean, if necessary, in accordance with Maintenance Procedure 4.3.3.3.

4.3.15.1—Evaporative Condenser

Weekly, inspect evaporative condenser. Service in accordance with Maintenance Procedure 4.3.15.0 and the following additional requirements:

Weekly
1. Check operating level in basin and adjust as necessary.
2. Check blow down valve and make-up water valves to make sure they're rking properly.

Monthly
1. Clean and flush basin (This may be required more often for towers located adjacent to highways, industrial sites, etc. with high particulate emissions).
2. Check water distribution system and sprays.
3. Check drift eliminators for proper position.
4. Blowdown evaporator water pump strainer.

Yearly
1. Clean and touch-up paint or other protective finish as necessary.
2. Dismantle and clean evaporator water pump strainer.

4.3.16.0—Pumps

Inspect and maintain pumps as follows:

Weekly
1. Perform visual inspection. Look for leaks, vibration, noise, etc. and service as needed. Check for coupler misalignment, coupler guards that are loose, etc.
2. Read pressure gauges and compare to normal conditions. If abnormal, investigate and correct problem. Check for air in the water system, indicated by little or no water flow, noise, or other abnormal operation. Vent and fill system as necessary.

Monthly
1. Lubricate pump bearings in accordance with manufacturer's written requirements.
2. Service motor in accordance with Maintenance Procedure 4.3.20.0.

3. For pumps with packing glands, check packing for wear and repack as necessary. If packing problems are consistent, replace packing glands with mechanical seals.

4. Check pump/motor alignment and condition of coupling. Adjust as needed. Replace coupling if there is wear.

5. Check mounting bolts. Adjust and tighten as needed.

Annually
1. Inspect bearings for wear. Using a dowel rod, listen to bearing for excess noise. Replace as needed.

2. Check motor bearings and temperature for condition.

Every 3 years
For motors over 50 nameplate horsepower, perform eddy current test.

4.3.16.1—Sump Pump
Inspect sump pump weekly and test for proper operation. Clean sump and inlet screen of debris.

4.3.16.2—Steam Condensate Pump
Every 6 months, steam condensate pumps should also be inspected and tested for proper operation

WHAT TO LOOK FOR	PROBABLE CAUSE	REMEDIAL ACTIONS
Pump will not start	Restriction in return line or pump suction prevents return of water.	Receiver inlet strainer requires cleaning. Closed return line gate valve. Low water pump cut-off, when used, may be open due to low water level.
Pump runs continuously	Excessive temperature causing pump to cavitate or become bound. temperature. Repair or replace steam traps.	Defective steam traps blowing through causing excessive condensate temperature. Repair of replace steam traps.

	Restricted flow	Check suction and discharge gate valves for proper setting. Check strainers for dirt. Check valve in discharge line, may be defective or installed backwards.
Pump short cycles or starts/stops in rapid succession.	Discharge line check valve defective, causing backflow.	Inspect check valve for tight closing and repair or replace if necessary.
Receiver flooding	Discharge valve opening.	Dirt in discharge bleed line.

4.3.17.0—Pneumatic Controls

Pneumatic controls require the following maintenance:

Air supply
Weekly
1.   Check compressor oil level and pressure

2.   Check for and repair air leaks.

3.   Inspect automatic drain for proper operation. Drain receiver tank if needed. Replace automatic drain if needed.

4.   Drain all air filter bowls.

Monthly
1.   Check percent on-time. If on-time exceeds 50%, inspect for air leaks, failed actuator diaphragm, etc. [If system has only one compressor, install a second, equally-sized compressor piped in parallel. Install check valve in each compressor's discharge line and valve so that either compressor can be removed for service without shutting down air supply. Install alternator control so compressors alternate operation.]

2.   Test safety relief valve(s). To test, adjust pressure reducing valve setpoint to just above relief valve setting. Relief valve should open. Reset pressure reducing valve to normal setting (20-25 psig). Relief valve should close. If the relief valve malfunctions, replace.

3.   Replace intake air filter to compressor.

4.   Check compressor drive belt wear, adjust belt tension.

5.   Inspect all valves for proper position, leaks, etc.

6.   Service compressor motor in accordance with Maintenance Procedure 4.3.20.0.

7.   Check discharge air filter(s) and pressure reducing valve. Replace as required.

8.   Check refrigerated air dryer(s). Measure discharge air for proper dewpoint temperature (-40°F). Repair or replace dryer as indicated. Clean condenser. [If the system has only one dryer, install a second, equally-sized dryer in parallel. Provide valves so that each dryer can be isolated and removed for service without shutting down air supply.]

Quarterly

1.   Replace compressor drive belt.

2.   Change compressor oil and oil filter.

3.   Test compressor on/off controller. Set pressure controller to 60 psig on pressure, 120 psig off pressure. Manually bleed air from receiver tank until pressure drops below 60 psig. Compressor should cycle on. When tank pressure rises to 120 psig, compressor should cycle off. If control malfunctions, replace controller.

Annually

1.   Inspect and tighten all electrical connections.

2.   Check motor amp draw.

Control Panels and Instruments
Monthly

1.   Calibrate transmitters and controllers. Adjust authority settings for dual input controllers.

2.   Test pressure gauges. Record gauge pressure reading, remove gauge, install calibrated test gauge and record reading. If readings

are within 10%, re-install original pressure gauge. If readings differ by more than 10%, replace original gauge with new.

Annually
1. Test all actuators for proper function. Re-build or replace faulty actuators and/or pilot positioners.

2. Test all control sequences by adjusting setpoints, high/low limit settings, etc. and check valves and dampers for proper operating range.

4.3.17.1—Electric/electronic Controls
Annually, electric/electronic controls require the following maintenance:
1. Room thermostats:
    a. Remove cover. Use squeeze bulb to blow out or low pressure compressed air to blow out dust and dirt. Clean contacts by hold them together lightly between the fingers and drawn a piece of hard finish paper, such as a business card, between them. *Never use emery cloth or abrasives.*
    b. Adjust setpoint(s) and heating or cooling anticipator as required by measuring the amperage with the system on and setting the anticipator to the measured amperage.
    c. If thermostat has a thermometer, compare the thermometer reading against ambient temperature based on an accurate digital thermometer reading. Adjust as required.

2. Electric actuator:
    a. Check and tighten all wiring connections.
    b. For damper actuators, check, adjust, and tighten linkage.
    c. Cycle actuator through its full operating range by adjusting setpoints, high/low limit settings, etc. Replace actuator that malfunctions.

4.3.17.2—Direct Digital Control Systems
Direct digital control systems require the following maintenance:

First Year (Warranty Period)
1. Operate all automatic valves and dampers through their full stoke. Ensure smooth operation through full stroke and appropriate seal-

ing or shutoff. Verify actuators are properly installed with adequate clearance. For pneumatic operators, verify spring range and adjust pilot positioners (where applicable).

2.   Verify calibration of sensors, sensor/transmitters, and sensor/controllers, including temperature, humidity, pressure, flow, and electricity consumption. Select at least 10% of the installed sensors, including at least one of each sensor type, for calibration testing. If calibration of 10% or more of this sample is found to be incorrect, select an additional 10% of the installed sensors for testing. *If calibration of 10% or more of this second sample is found to be incorrect, test all sensors.*

Utilize the following calibration test procedure for air, steam, and water temperature, air humidity, air static pressure, and steam pressure with DDC system sensors in place:

a.   Place calibration sensor in the controlled medium (air, water, or steam). Caution shall be exercised that DDC system and calibration sensors do not interfere with each other (such as for temperature or pressure sensing) and that the calibration sensor does sense the true conditions.

b.   Wait sufficient time for the calibration sensor to stabilize. The time required depends on the time constant and the initial condition of the sensor.

c.   Coordinate time between the site and central control room for taking readings.

d.   Simultaneously take readings of the calibration instrument and the DDC system readout terminal.

e.   Take at least one more reading for the same sensor. The range of the two readings taken shall be as wide as the HVAC operation allows, but not more than that called for in the contract specifications. In order to have the wide range desired the second test may need to be conducted at a different time from the first test when the HVAC equipment experiences different loads.

f.   Compare readings from the calibration instrument and the DDC system. They must be within the accuracy requirements of the specifications. If not, replace them.

3.   Verify the existence and operation of specified software.

a.   Verify backup system operations and switchovers including redundant processors, backup power supplies, battery backed

memories, etc.
b. Verify DDC system command software. This may be checked by issuing commands at the operator's console and observing display, printer output, or HVAC equipment responses.
c. Test software for checking input commands and issuing error messages. Enter various correct and incorrect commands and monitor results
d. Check system and point addressing. Enter command to display I/O data. Verify all data points defined on the drawings and/or required by the specifications.
e. Test start-stop or enable-disable of HVAC equipment or DDC system control points. Enter commands to start/stop selected HVAC equipment, and to disable and enable selected points.
f. Test operator override/automatic mode. Enter command to change selected automatic control under DDC system to manual and vice versa.
g. Test display format. Enter commands to display data and graphics on terminal and graphic display. Check display content for adequacy and clarity as specified.
h. Test ability to modify, cancel and confirm operator's commands. Verify by entering commands.
i. Test setpoint adjustment and limiting. Enter commands to adjust set points of controllers and range limits of the controlled media. Verify by display. Also enter commands to adjust setpoints outside their range limits. DDC system shall display error messages.
j. Test system access and access level control. Try to log on to system with both incorrect and correct ID codes. Try to enter different commands with different access level of the operators. The responses of the DDC system shall be as specified.
k. Start/stop equipment. Enter command to start or stop selected equipment. Also reset time to initiate automatic mode. Verify responses by observation of equipment and DDC system display.
l. Change parameter of points. Enter commands to change parameters of selected points such as high and low limit alarms, scale factor, etc. to test the adequacy of software.

4. Verify graphic display of each HVAC system and component. Confirm that the graphic is in accordance with the design data and re-

viewed submittals, includes all data points required, displayed data is correct and in the correct format and units, and changes in point conditions or status are accurately updated. Evaluate the refresh rate of data display.

5. Verify report generating (status, profile, energy, etc.) by entering commands to generate reports such as all points, trend, total display of a system, timed display, and other specified reports. Examine the report content for general format, system/point code, time interval of reporting, point status/value/unit, energy amount/rate/ unit, status of control and set time (manual or automatic), and other specification required information.

6. Test alarm reporting by initiating alarm conditions of different points at different alarm levels in sequence to examine alarm reports. The reports shall show alarm location and device, alarm time, cause of alarm, current status of the point, etc. as required in the specifications. When alarm conditions are removed the printer shall print updated status report. Also verify audible alarm operations in accordance with specification requirements. Then initiate alarm conditions at different levels at the same time to check alarm priority.

7. Test DDC system application software via the following test procedures:
   a. Scheduled start-stop control. Verify that input includes start/ stop time and days for specified equipment. Verify input for time delays of specified equipment. Check holiday effect. For each air handler, log operation for at least three days to confirm that unit starts and stops in accordance with the schedule.
   b. Optimum start-stop control. Verify that the required input includes space temperature, outdoor temperature, occupied time and days of the week, and the response time of the air handling equipment, on an individual system by system basis. For each air handler, log space temperature for at least three days to confirm that space is at required temperature by the start of the occupied period.
   c. Electrical demand limiting control. Verify required input data, as follows: electrical loads under control, load priority, demand

metering interval, demand limit setpoint, delay time, minimum off-time, minimum on-time, and maximum off-time. Test control by the following process:

Override demand sensor and input demand limit setpoint 10% lower than current electrical consumption rate (kW).

Confirm equipment shutdown by observation of equipment and/or display.

d. Day-night setback control. Confirm that required input data is provided. To test control, change the setback time from occupied to unoccupied time and confirm that HVAC systems respond to the setback mode. If system is an air-handling system, the outside air damper should close and the fan should cycle to maintain the setback temperature setpoint.

Change the setback temperature setpoint to 5°F higher than the actual space temperature. The system should operate to increase the space temperature to the new setpoint condition. For air-handlers, the outside air damper should remain closed.

e. Economizer cycle control. If the outdoor air temperature is above the discharge air temperature setpoint, but below the defined 100% outdoor air changeover temperature setpoint, verify that the outside air damper is fully open, the return air damper is fully closed, and that the air-handler discharge air temperature is maintained by modulation of the chilled water control valve.

f. Heating/cooling coil discharge temperature reset. Verify that the required input points are provided and that the coil discharge air temperature setpoint is in accordance with the input conditions and the sequence of operation defined on the control drawings.

g. Hot water temperature reset. Verify that the required input points are provided and that the heat exchanger supply hot water temperature setpoint is in accordance with the input conditions and the sequence of operation defined on the control drawings.

h. Chilled water temperature reset. Verify that the required input points are provided and that the chilled water supply temperature setpoint is in accordance with the input conditions and the sequence of operation defined on the control drawings.

i. Condenser water temperature control. Verify that the required input points are provided and that the condenser water supply temperature setpoint is in accordance with the input conditions and the sequence of operation defined on the control drawings.

Year 5: "Re-commission" system by repeating all checks and tests done during the first year.

Year 10: The system has become both technically and functionally obsolete and should be replaced.

4.3.18.0—Water Piping

Annually, visually inspect all water piping exposed to view for leaks, corrosion, and condition. Inspect insulation in accordance with Maintenance Procedure 4.3.19.0. Cycle valves to make sure they operate. Make repairs as needed.

4.3.18.1—Steam & Condensate Piping

Annually, visually inspect all steam and condensate piping exposed to view for leaks, corrosion, and condition. Inspect insulation in accordance with Maintenance Procedure 4.3.19.0. Cycle valves to make sure they operate. Make repairs as needed.

4.3.18.3.1—Expansion/compression Tank

Monthly, check expansion/compression tank sight glass for water level. If water-logged, drain down and pressurize system to create an air cushion.

4.3.18.3.2—Piping Vibration Isolator

Monthly, visually inspect piping vibration isolator for leaks or failure. Replace as required.

4.3.18.3.3—Strainers

Inspect and service strainers as follows:
1. Monthly, inspect for leaks and blow down.
2. Annually, remove, clean, and reinstall strainer element or basket.

4.3.18.3.4—Rotary Air Separator

Inspect and service rotary air separator as follows:
1. Monthly, inspect for leaks and blow down (if separator has integral

strainer). [Note: If the system pumps are equipped with Y-type or basket strainers, remove the integral strainer to reduce maintenance requirements.]

2.   Monthly, check make-up water connection. Check pressure gauges for proper water pressure reducing valve operation.

3.   Annually, remove, clean, and reinstall strainer element (if separator has integral strainer).

4.3.18.3.5—Water Pressure Relief Valve
Every 6 months, examine water pressure relief valve for leaks and/or corrosion and test for proper operation by raising lever or stem and then releasing it. If valve is damaged or malfunctions, replace.

4.3.18.4.1—Steam Pressure Reducing Valve
Every 6 months (or beginning of each heating season), each steam pressure reducing valve (PRV) should be inspected for proper operation, as follows:

1.   Joint leaks: *Do not allow a leak to continue; tighten bolts, replace gaskets or packing as necessary.*

2.   Proper control of pressure and/or temperature: Check to see that the operating conditions initially adjusted are still set. (An additional check is to alter the adjustment slighting to see if the system responds.)

3.   Proper solenoid pilot actuation: Make sure it is functioning properly.

4.   Blowdown system strainers: If excessively dirty, blowdown should be done once a week.

5.   Inspect for dirt collected at bleedpoint and restriction elbow.

When a PRV does not perform satisfactorily, it is best to determine the cause by making the easiest inspections first before disassembling. The following list of possible causes are arranged in order of difficulty.

WHAT TO LOOK FOR	PROBABLE CAUSE	REMEDIAL ACTIONS
Delivery Pressure is Low or Drops Off	Low inlet pressure.	Fully open supply valve. Unclog strainer. Check for low boiler output or upsteam blockage and make necessary correction.
		Feedback control line valve closed. Solution: Fully open supply valve.
	Pilot valve range incorrect or adjustment altered.	Check nameplate for operating range of pilot. Change pilots if desired operating range is beyond that on existing pilot. Readjust to desired operating condition.
	Bleed orifice missing. If a straight fitting was installed in place of the bleed orifice or if it is worn, the pilot signal may not fully open the main valve.	Install a new bleed orifice.
	Pilot valve lines blocked.	Remove the line to the main valve and downstream feedback line. If they are not clear, replace.
	Pilot valve or main valve malfunction.	Repair valve.
Delivery pressure is high or overrides setpoint.	Open valve on by-pass line.	Close the valve.
	Pilot adjustment altered.	Readjust to desired operating condition.
	Pilot valve lines or bleed orifice blocked.	Remove, check and replace as required.

| Pilot valve or main valve malfunction. | Repair valve. |
| Pressure drop limits exceeded. (Recommended maximum single stage reduction is 100 psig.) | Reduce the pressure drop. |

4.3.18.4.2—Steam Safety Relief Valve

Test steam safety relief valve by increasing steam supply pressure, by changing pressure reducing valve setting, to the relief valve setting. Valve should open. Reduce supply pressure. Valve should close. If valve does not open or close properly, replace valve. *Do not attempt to repair valve.*

4.3.19.0—Piping Insulation

Annually, inspect condition of insulation on piping and equipment.

1. Inspect for missing insulation and damaged (tears, holes, abrasion, etc.) jacket(s). Replace insulation and/or jackets as required.

2. Hand-squeeze insulation on fittings for proper fill. If insulation is soft or it feels as if the PVC fitting cover is not "filled," remove PVC fitting cover, install correctly-sized pre-formed fitting insulation, and re-install PVC fitting cover.

3. Cover ASJ exposed to view with PVC or aluminum outer jacket to prevent mechanical damage. (Use PVC for chilled water piping insulation.)

4. For fiberglass insulation on chilled water systems, examine for mold growth on the jacket, especially at or near fittings. Determine the extent of wet or moist insulation. If found, wait until heating season and remove and replace with cellular glass.

5. Check calcium silicate insulation on steam piping and equipment for erosion, cracking or failure of plaster outer coatings, etc. Repair or replace as necessary.

6.   Check condition of flexible elastomeric insulation for deterioration, especially if installed outdoors (typically on refrigerant lines). Replace as required.

7.   Check outdoor fiberglass or calcium silicate insulation for water intrusion. If outdoor PVC or aluminum jacket is missing, damaged, or corroded, replace underlying insulation and outer jacket with new.

4.3.20.0—Motors
Motor maintenance is required, as follows:

1.   Lubricate electric motors in accordance with the following schedules:

Open, drip-proof (ODP) HVAC electric motors:
     Re-lubricate on the basis of the following schedule:

Motor RPM	Ambient Temperature (F)	Regreasing Frequency
1800	100 or less	6 months
	150	3 months
	200	1 month
	250	2 weeks
3600	100 or less	4 months
	150	2 months
	200	3 weeks
	250	10 days

Lubricant should comply with the following:

Motor RPM	Temperature (°F) Ambient	Temperature (°F) Operating	Grease Recommendation
Up to 3600	-20	250	Lithium grease or poly-urea grease
Over 3600	0	200	Lithium complex grease with light viscosity oil or non-soap grease, like

			MIL-G-81422 grease
Below 3600, High Temp	—	400	Lithium complex grease, High Temp polyurea grease, or Mil-G-81322 grease
Vertical Motors Up to 3600	−20	250	Lithium base grease with Up to 3600 heaver than NLGI 2 consistency

Motor bearings must be flushed if new grease is not compatible with the old grease or the if the old grease has become contaminated. When flushing, make sure that motor windings do not become contaminated.

TEAO (Totally Enclosed, Air Over) motors
   a.   Re-lubrication is not required, they have permanently lubricated ball bearings and special moisture protection on the bearings, shaft, and windings.
   b.   Quarterly, clean motor surfaces must be cleaned to maintain proper motor cooling.

TEFC (Totally Enclosed, Fan Cooled) motors
   a.   Re-lubricate in accordance with schedule for ODP motors.
   b.   Every 3 months, clean fan and "weep holes" at the bottom of the end housings that allows condensation or any other moisture accumulation to drain.

2.   Annually, inspect each motor
   a.   Check control box for dirt, debris, and/or loose terminal connections. Clean and tighten connections as needed.
   b.   Check motor contactor for pitting or other signs of damage. Repair or replace as needed.

4.3.21.0—Closed Systems Water Treatment
   Water treatment for chilled water and hot water systems, which are "closed recirculating water" systems, is simple, since closed systems have many advantages
1.   There is no loss of water in the system (except when a leak occurs)

and, thus, no need for make-up water. Therefore, deposition or scaling is not a problem.

2. Once filled and entrained air removed, the closed system creates an anaerobic environment that eliminates biological fouling as a problem.

3. Closed systems reduce corrosion problems because the water is not continuously saturated with oxygen (an oxidizer) as in open systems. The low temperatures common to HVAC systems further reduces the potential for corrosion.

Water treatment to eliminate the potential for corrosion is required only when the water system is initially filled with water, or when it is drained and re-filled due to maintenance, repair, or modification.

Corrosion in a closed system can occur due to oxygen pitting, galvanic action and/or crevice attack. To prevent these conditions, the "shot feed" method of chemical treatment is used. With this method, a bypass chemical feeder is used to add treatment chemicals to the system in a one-time "shot" just after the system is filled with water. For the mixed metallurgy (steel and copper) systems typical for HVAC water systems, a molybdate corrosion inhibitor is best, with treatment limits of 200-300 ppm recommended.

4.3.21.1—Condenser Water Treatment
Water treatment programs for condenser water systems is required.

1. **Deposition Control:** *Deposition* is the term that is used to describe scale and other deposits that may form on the heat transfer surfaces of a condenser water system. Water has been called the "universal solvent" and make-up water will always contain dissolved mineral solids to a greater or lesser extent. The purpose of the deposition control portion of a condenser water treatment program is to prevent amount of dissolved solids from becoming so high that they will begin to be deposited on the wetted surfaces of the system.

Deposits of calcium carbonate on wetted surfaces in a condenser water system act as insulation and seriously reduce the heat transfer effectiveness. Deposit thickness can increase to where it reduces flow areas and, thus, significantly reducing flow rates as associated velocity and pressure loss increases. In effect, significant calcium carbonate deposits will destroy the effectiveness of the condenser water system and the refrigeration system it supports.

As a portion of the condenser water flow is lost by evaporation through the cooling tower, the concentration of dissolved solids increases. Make-up water, which is added to the condenser water system to offset evaporation losses, will add dissolved minerals at a lower concentration level than the condenser water and, thus, some equilibrium concentration level will be maintained. However, if this equilibrium concentration level is high enough that deposition occurs, then a program to control solids concentration at a lower level is required.

The concentration of dissolved solids can be reduced by adding more make-up water, which has a lower concentration, to the condenser water. However, to add water to the system, an equal amount must be removed from the system by *blowdown*, the intentional "dumping" of condenser water to drain.

The tendency to form calcium carbonate deposits (scale) is a function of the temperature, pH, calcium hardness, total alkalinity, and total dissolved solids of the water.

Temperature: The condenser water supply temperature for HVAC systems will range from 70°F at low ambient wet bulb and light load to 85°F at design wet bulb temperature and load. Most scale-forming dissolved solids have the unusual property of becoming less soluble as the water temperature increases. Thus, the greatest potential for scaling occurs at design conditions.

pH: Water is made up of hydrogen and oxygen as follows:

H^+ protonated water molecule

OH^- hydroxal radical or hydroxide

Combining H^+ with OH^- yields H_2O—water.

Chemically, pH is a negative logarithmic scale defining the relative concentration of H^+. Each increase or decrease of pH by 1, represents a 100-fold increase or decrease in the amount of acidity or alkalinity (base).

Calcium Hardness: The amount of calcium carbonate in the water, ppm.

Total Alkalinity: The amount of negatively charged ions of hydroxide (OH^-), bicarbonate (HCO_3^-), and carbonate (CO_2^{2-}) in the water, ppm, and represents the ability of the water to neutralize acid. Alkalinity occurs in all water with a pH above 4.4.

Total Dissolved Solids: The amount of all dissolved solids (calcium, magnesium, phosphate, iron, etc.) in the water.

Water hardness and alkalinity is a function of the hardness and alkalinity of make-up water, the amount of evaporation and drift loss from the cooling tower operation, and the blowdown to proposed to yield the desirable pH to prevent both deposition and corrosion. The term *cycles of concentration* defines the ratio of the desired concentration of dissolved solids in the condenser water to the concentration of dissolved solids in the make-up water, as follows:

Cycles (of Concentration) =

[Dissolved Solids (ppm) in Blowdown]/[Dissolved Solids (ppm) in Make-up]

This relation can be expressed in terms of water flow as follows:

Cycles (of Concentration) = MU/BD

where

    MU = total make-up water flow, which is the sum of evaporation + blowdown, gpm

    BD = blowdown flow, gpm

Since the amount of evaporation, E, is typically 0.1% of the condenser water flow rate per degree of range, MU in the equation above can be replaced with the value (E + BD). above can then be rearranged to yield the following equation

    BD = E/(Cycles – 1)

Thus, after the number of cycles is determined based on the make-up and desired condenser water (blowdown) concentration of dissolved solids, the actual blowdown requirement can be calculated.

Since drift water loss is not included in these calculations, the actual required BD flow can be reduced by the amount of drift loss from the tower.

Controlled blowdown is based on continuous monitoring of the water hardness as indicated by its conductivity. Automatic control minimizes the waste of water and water treatment chemicals and is the preferred method.

The amount of make-up water is reduced significantly as the number of cycles is increased from 2 to 6. However, there is only a further 5% reduction as the cycles is increased from 6 to 10, and only a further 2% reduction as cycles is increased to 20. *Therefore, in most cooling tower applications, cycles of concentration is maintained between 5 and 10 and deposition inhibitors are added as necessary.* While lower cycles represent loss of more water and treatment chemicals, the amount of treatment chemicals required tends to go down with cycles, and 5-10 cycles usually represents a good balance point.

For a dissolved salt to precipitate and deposit on the wetted metal surfaces of the condenser water system, it forms a crystal growth that attaches itself to metal surfaces as it comes out of solution. The most common deposition inhibitors used in condenser water systems are *phosphonates*, which are organic phosphate compounds, such as HEDP, which function by adsorption on the crystals as they form and prevent them from attaching to metal. Thus, these crystals precipitate out of solution, usually in the tower basin.

2.   **Corrosion Control**: For corrosion control, condenser water pH should be maintained between 4 and 10. In most cooling tower water treatment programs, the desirable range for pH is between 8 and 9 in order to limit water alkalinity to a reasonable level (400 ppm or less). However, in metal towers, a pH of 7.0-8.0 is preferred to help prevent white rust corrosion. Therefore, at least for most metal cooling towers, the ideal pH range is 7.5-8.5.

Metal corrosion occurs as a result of *galvanic action* at a negatively charged "pole" or site on the metal surface. Corrosion, then, is the loss of metal; it literally "dissolves." Corrosion can exhibit two characteristics depending on the underlying reason for the anodic and cathodic sites.

General corrosion is wide spread and is caused, usually, by impurities in the metal or characteristics of the metal or its environment that results in an overall fouling of the metal surface. *Localized corrosion* results, mostly, from scratches, stress, or localized environment and the most common reason for "metal failure."

If dissimilar metals with different electrical potentials are used in a condenser water system, galvanic corrosion is enhanced and the metals simply corrode faster, particularly at and near the point(s) of contract between the metals.

The first step in corrosion control is to minimize the contact between water and mild steel materials. Ideally, all primary wetted surfaces—wet decks for induced draft towers and basins— should be constructed of stainless steel. The use of plastics or fiberglass for the tower casing, wet deck covers, intake louvers, drift eliminators, and fill also reduces the potential for corrosion. Mild steel used for the tower structural frame, etc. should be galvanized and coated with an epoxy or polymer final protective coating.

Piping in most condenser water systems is steel and must be protected from corrosion. This accomplished by using one or more treatment programs, as follows:

a.   **Passivating (Anodic) Inhibitors**: These chemicals form a protective oxide film on the metal surface which is not only tough, but, when damaged, quickly repairs itself. Typical chemicals that act as passivating inhibitors include molybdate, polyphosphates, and orthophosphate. These chemicals are oxidizers that promote passivation by increasing the electrical potential of the iron. The drawback to the use of molybdate is its expense, and it is used only when blowdown levels are kept as low as possible. *Othophosphates should not be used in condenser water systems containing stainless steel since it will make the metal brittle over time.*

b.   **Precipitating Inhibitors**: At cathodic sites, the localized pH at the site is increased due to the higher concentration of hydroxide ions that are being produced. Precipitating inhibitors form complexes that are insoluble at the higher pH and, thus, precipitate out of the water. Zinc is good precipitating inhibitor. Molybdate will also act as a precipitating inhibitor and, thus, can serve as a corrosion inhibitor using two mechanisms.

c.   **Adsorption Inhibitors**: These are organic compounds containing
     nitrogen, such as amines, or sulfur or hydroxyl groups. Due to the
     shape, size, orientation, and electrical charge of the molecule, they
     will attach to the surface of the metal, preventing corrosion. Their
     drawback is that they form thick, oily surface films that reduce heat
     transfer capability.

Each program is designed for particular condenser water pH range
and water chemistry as follows:

Treatment Program	Water pH and Chemistry
Zinc	7.5-8.5
Molybdate or Molybdate/Zinc	7.5-9.5
Orthophosphate	7.5-8.5, with a phosphate deposition inhibitor
Organic Adsorption	7.5-9.5, with 300-500 ppm alkalinity

To protect copper in heat exchanger tubes and piping from corrosion, aromatic triazoles, such as benzotrizole (BZT) and tolyltriazone
(TTA), are used. These compounds bond with the cuprous oxide on the
metal surface and protect it.

Two additional special conditions of corrosion must also be addressed by the water treatment program:

a.   **Microbiologically Induced Corrosion (MIC)**: MIC is caused by sul-
     fate-reducing bacteria (SRB) in the water and is usually evidenced
     by reddish or yellowish nodules on metal surfaces. When these
     nodules are broken, black corrosion by-products are exposed and
     a bright silver pit is left in the metal. A "rotten egg" smell when the
     nodule is broken is also evidence of SRB corrosion.

     SRBs obtain their energy from the anaerobic reduction of sul-
     fates that are available in most water. The bacteria contains an en-
     zyme that enables it to use hydrogen generated at a cathodic site to
     reduce sulfate to hydrogen sulfate and act like a cathodic "depolar-
     izing agent." *Iron corrosion by this process is very rapid and, unlike rust-
     ing, is not self-limiting.*

     Once SRBs begin to grow in a system, they are very difficult to
     eliminate. Thus, a *preventative program* is far more effective that an
     clean-up program.

1) Keep the system clean through sidestream filtration and regular cleaning.
2) Prevent contamination by oils and/or grease. Even very small amounts can cause problems.
3) Eliminate potential bacteria sources (such as bathroom and kitchen vents, diesel exhaust, etc.) near cooling towers.
4) Run the condenser water pump as much as possible. Do not allow stagnant conditions to exist.

The condenser water system should be regularly tested for SRBs in accordance with ASTM Standard D4412-84 *Standard Test Methods for Sulfate-Reducing Bacteria in Water and Water-Formed Deposits*. Normally, control of SRBs can be accomplished with the housekeeping measures outlined above, coupled with the use of an oxidizing antimicrobial. *However, if SRBs do become established in the system, biocides are no longer effective and a special clean-out program must be designed for each system.*

b.   **White Rust**: The term *white rust* refers to the premature, rapid loss of galvanized coating on cooling tower metal surfaces. White rust is evidenced by a white, waxy, non-protective zinc corrosion deposit on wetted galvanized surfaces. This rapid loss of the galvanizing results in the corrosion of the underlying steel and equipment will have drastically shortened performance life.
White rust may form if the following conditions exist:
1) The galvanized coating is not properly "passivated" when the tower is placed in service. (Passivation is a process that allows the zinc coating to develop a natural nonporous surface of basic zinc carbonate. This chemical barrier prevents rapid corrosion of the zinc coating from the environment, as well as from normal cooling tower operation. The basic zinc carbonate barrier will form on galvanized surfaces within eight (8) weeks of tower operation with water of neutral pH (6.5-8.0), calcium hardness of 100-300 ppm, and alkalinity of 100-300 ppm.)
2) Condenser water is maintained at pH above 8.0.
3) High condenser water alkalinity (above 300 ppm).
4) Low condenser water calcium hardness level (below 100 ppm).
5) The lack of phosphate-based corrosion inhibitor in the condenser water treatment program.

For most galvanized metal HVAC cooling towers, white rust will occur if not prevented by the following maintenance steps:

1) Provide a secondary barrier coating on all wetted surfaces, such as epoxy or polymer finish for a new tower or coaltar (bitumen) on an existing tower. [An even better approach is to specify new towers to have wetted surfaces such as basins and wet decks to be constructed of stainless steel. This option is normally available for only a 113-20% cost increase.]

2) Maintain water pH between 7.0 and 8.0, which may require pH control.

3) Make sure the galvanized tower is properly passivated upon system start-up. Where white rust has occurred, the metal can be "re-passivated" by treating the surface with a 5% sodium dichromate 0.1% sulfuric acid, brushing with a stiff wire brush for at least 30 seconds, then rinsing with thoroughly.

4) Incorporate a phosphate-based product into the water treatment program, along proper dispersants.

3. **Biological Fouling**: Biological fouling results from bacteria, fungi, zooplankton, and phytoplankton or algae introduced through make-up water or filtered from the air passing through an HVAC cooling tower. "Fouling" results when these micoorganisms grow in open systems rich in oxygen (an aerobic process) and form slime on the surfaces of the tower, piping, and heat transfer surfaces of the condenser water system. Slime is an aggregate of both biological and non-biological materials. The biological component, called the *biofilm*, consists of microbial cells and their byproducts. The non-biological components consist of organic and/ or inorganic debris in the water that has become adsorbed or imbedded in the biofilm layer.

The impact of biological fouling is twofold: the slime acts as an insulator and reduces heat transfer efficiency in the system and microbial activity within the slime can accelerate corrosion by creating a localized oxygen-rich environment that accelerates oxidation.

The best method of controlling biological fouling is to keep cooling towers clean, as defined in Maintenance Procedure 4.3.14.0 Then, the use of chemical treatment will complete the control chore.

There are two kinds of antimicrobial chemicals or *biocides* used in cooling tower water treatment programs to control biological fouling: oxidizing and nonoxidizing.

a.   **Oxidizing chemicals** include chlorine, bromine, and ozone that oxidize or accept electrons from other chemical compounds. Used as antimicrobials, these chemicals reacted directly with the microbes and degrade cellular structure and/or deactivate internal enzyme systems. They penetrate the cell wall and disrupt the cell metabolic system to "kill" it. (Warning! Oxidizing chemicals, particularly chlorine, can react with steel, including stainless steel, and cause rapid corrosion. To prevent this, concentrations of these chemicals must be kept low, ideally to less than 0.7 ppm. Oxidizing chemicals must be introduced into the condenser water system is such a way to be rapidly dispersed to prevent localized high concentrations.)

b.   **Nonoxidizing antimicrobials** attack cells and damage the cell membrane or the biochemical production or use of energy by the cell, resulting in its death, and are sometimes referred to as "surface-active" biocides. Typical nonoxidizing biocides include isothiazolines, gluteraldehyde, MBT, and polyquat.

Microbials in condenser water systems can become resistant to a single method of attack, or some microbials may be more or less immune to one type of attack. Therefore, it is recommended that both types of treatment chemicals be used (oxidizing and nonoxidizing), either blended together or in alternating treatment patterns, as indicated by periodic water testing results. The key to a successful biological treatment program is maintaining adequate chemical treatment levels at all times via continuous feed of antimicrobials into the condenser water system.

4.   **Treatment for Wooden Cooling Towers**: Wood is composed of three major elements: cellulose, which forms long fibers in wood and give it its strength; Lignin, the soft material that acts as a cementing agent for the cellulose; and Extractives, the natural compounds that enable wood to resist decay.

Wood deterioration can result from chemical attack, biological attack, and/or physical decay due to (elevated) temperature degradation or rupture of wood cells by crystallization of dissolved solids in the condenser water. Physical decay is relatively uncommon in HVAC applications.

Chemical attack commonly results in *delignification* of the wood, resulting in wood that has a white or bleached appearance and a fibrillated surface. Delignification is usually caused by oxidizing agents and alkaline materials and is particularly severe when high chlorine residuals (more than 1 ppm free chlorine) and high alkalinity (pH of 8 or higher) occur simultaneously.

The biological organisms that attack cooling tower wood are those that use cellulose as a source of carbon for growth. These organisms degrade the cellulose by secreting enzymes that convert the cellulose into compounds they can then absorb as food. This type of attack deletes the cellulose and leaves a residue of lignin. The wood appears dark, loses much of its strength, and becomes soft and spongy.

Biological attack is of two basic types: surface rot and internal decay. Surface rot is easily detected and can be repaired, but internal decay, usually found in the plenum area, cell partitions, decks, fan housing, supports, and other areas that are not continuously flooded, is more difficult to detect simply because the exterior of the wood will display little or no sign of the rot.

Water treatment and preventative maintenance is the only way to prevent wood deterioration:

a. For flooded areas of the tower, use nonoxidizing antimicrobials to control slime and prevent biological attack.

b. Internal decay is the most common problem for non-flooded areas of a tower and these areas should be thoroughly inspected at least twice each year. Test for soundness with a blunt probe (long, thin screwdriver) and, when suspect, a sample of wood should be examined microscopically to detect internal microorganisms. *Any infected wood should be replaced immediately* with pretreated wood to prevent the spread of the infection to healthy wood. (Periodic spraying with an antifungal treatment may be helpful in reducing biological attack, but it must be done very thoroughly and on a regular schedule, otherwise it is a waste of time and chemicals.)

5. Control of *Legionella*: Because both cooling towers and evaporative condensers use a fan system to move air through a recirculated water system, they introduce a considerable amount of water vapor into the surroundings, even with drift eliminators designed to limit vapor release. In addition, this water is typically in the 65°-125°F range, ideal for *L. pneumophila* growth.

The key maintenance issues for control of *Legionella* are as follows:

a. Eliminate stagnant water areas.

b. Eliminate controllable sources of nutrient to the cooling water system.

c. Maintain overall system cleanliness and provide good biological control.

d. Use the best technology to maintain the lowest possible drift rate and maintain drift eliminators to ensure their ongoing effectiveness.

Specific water treatment measurements required for prevention of *Legionella* occurrence are as follows:

a. First, test for *Legionella* in cooling tower water on a every 6 months. While there is an academic debate over the cost-vs-benefit of routinely testing for *Legionella*, it is foolish (from both an ethical and a liability point of view) to ignore any potentially life-threatening condition in a facility.

Contact a laboratory experienced in performing *Legionella* analyses on environmental samples using the CDC "Standard Culture Method." Also, concurrent sampling should be performed on the bulkwater and surface deposits for microscopic detection of higher life forms, along with total aerobic heterotrophic counts. Collect. bulk water samples from several locations within the system (e.g., makeup water, hot return water, basin water, and from sample taps on heat exchangers remote from the cooling tower if available). Where evident, collect deposit samples from the basin walls, tower fill, and distribution decks.

b. Test results may indicate the following:

1) A low *Legionella* count with an undetectable or small population of amoebae/protozoa (higher life forms) and low biofilm counts (low bacteria numbers) is a good indication of a clean, well maintained system with low risk to health.

2) A low bulk water *Legionella* count along with low numbers of higher life forms in deposits, but with high biofilm counts may indicate a low present health risk but suggests the potential for future problems if steps are not taken to reduce biofilm levels. Since protozoa that promote *Legionella* amplification graze on bacteria in biofilms, the presence of significant biofilm can promote the development of higher, and thus potentially more dangerous, levels of *Legionella*.

3) A low bulk water *Legionella* count associated with a large number of higher life forms indicates a strong potential for amplification and the low *Legionella* count cannot therefore be interpreted to indicate a system with a low health risk.

c.   Provide continuous halogenation treatment (chlorine or bromine):

1) For relatively clean systems or where clean potable water makeup is used, feed a source of halogen (chlorine or bromine) continuously and maintain a free residual. Continuous free residuals of 0.5 to 1.0 ppm in the cooling tower hot return water are recommended by many agencies. Periodic monitoring of the residual at sample points throughout the cooling water system is needed to insure adequate distribution. The effectiveness of either halogen decreases with increasing pH; bromine is relatively more effective at a higher pH (8.5 to 9.0).

2) Stabilized halogen products should be added according to the label instructions, and sufficient to maintain a measurable halogen residual.

3) A biodispersant/biodetergent may aid in the penetration, removal, and dispersion of biofilm and often increases the efficacy of the biocide.

4) Continuous halogen programs may require periodic use of nonoxidizing biocides. These may be required to control biofilm and planktonic organisms in systems that use makeup water from other than potable water sources, and those with process leaks or contamination. The choice of nonoxidizing biocides should be based on the results of toxicant evaluations. Reapply as dictated by results of biomonitoring.

d.   Hyperhalogenation as practiced is the maintenance of a minimum of 5 ppm free halogen residual for at least 6 hours. Periodic on-line disinfection may be necessary for systems

1) That have process leaks,

2) That have heavy biofouling,

3) That use reclaimed wastewater as makeup,

4) That have been stagnant for a long time,

5) When the total aerobic bacteria counts regularly exceed 100,000 CFU/ml, or

6) When *Legionella* test results show greater than 100 CFU/ml

Periodic hyperhalogenation will discourage development of large populations of *Legionella* and their host organisms. Consequently, periodic hyperhalogenation may eliminate the need for conducting more complicated and higher risk off-line emergency disinfection procedures.

e. The following emergency disinfection procedure is based on OSHA and other governmental recommendations and is required when very high *Legionella* counts exist (i.e., >1000 CFU/ml), in cases where Legionnaires disease are known or suspected and may be associated with the cooling tower, or when very high total microbial counts (>100,000 CFU/mL) reappear within 24 hours of a routine disinfection (hyperhalogenation)

1) Remove heat load from the cooling system, if possible.

2) Shut off fans associated with the cooling equipment.

3) Shut off the system blowdown. Keep makeup water valves open and operating.

4) Close building air intake vents in the vicinity of the cooling tower (especially those downwind) until after the cleaning procedure is complete.

5) Continue to operate the recirculating water pumps.

6) Add a biocide sufficient to achieve 25 to 50 ppm of free residual halogen.

7) Add an appropriate biodispersant (and antifoam if needed).

8) Maintain 10 ppm free residual halogen for 24 hours. Add more biocide as needed to maintain the 10 ppm residual.

9) Monitor the system pH. Since the rate of halogen disinfection slows at higher pH values, acid may be added, and/or cycles reduced in order to achieve and maintain a pH of less than 8.0 (for chlorine-based biocides) or 8.5 (for bromine-based biocides).

10) Drain the system to a sanitary sewer. If the unit discharges to a surface water under a permit, dehalogenation will be needed.

11) Refill the system and repeat steps 1 through 10, above.

12) Inspect after the second drain-off. If a biofilm is evident, repeat the procedure.

13) When no biofilm is obvious, mechanically clean the tower fill, tower supports, cell partitions, and sump. Workers engaged in tower cleaning must wear (as a minimum) eye protection and a ½ face respirator with High Efficiency Particulate (HEPA) fil-

ters, or other filter capable of removing >1 micron particles.

14) Refill and recharge the system to achieve a 10 ppm free halogen residual. Hold this residual for one hour and then drain the system until free of turbidity.

15) Refill the system and charge with appropriate corrosion and deposit control chemicals, reestablish normal biocontrol residuals and put the cooling tower back into service.

4.3.21.2—Water Treatment for Steam Systems

Water treatment programs for steam systems is required to reduce or eliminate a wide range of steam system failures:

1.   Poor heat transfer and/or overheating and rupture due to deposits on tubes.

2.   Corrosion failures due to oxygen pitting, chelant corrosion due to excess concentration of sodium salt over a period of time, caustic attack due to tube deposits in phosphate treated boilers caused by caustic dissolving magnetite film, and/or acid attack due to poor pH control.

To avoid these potential problems in steam boiler systems, water treatment systems and programs must be implemented, as follows:

1.   **Make-up Water Pretreatment:** The most common method of reducing water hardness and alkalinity is to "soften" the incoming make-water to reduce the concentration levels of dissolved sodium salts via a *zeolite water softening process*. This is an *Ion exchange process* that uses "strong acid cation" resin to exchange calcium and magnesium ions for sodium ions. Regeneration of resin is done by treating with sodium chloride solution, then rinsing.

Deaeration, the next step in pretreatment, is designed to remove dissolved oxygen from make-up water produces localized corrosion (pitting) in piping and boiler tubes. Dissolved gases can be removed by lowering the pressure in the atmosphere contacting the liquid by "vacuum deaeration." However, this process is relatively inefficient so "pressure deaeration" is normally used for boiler feedwater pretreatment. With this process, feedwater is sprayed into a low pressure (typically 5 psig) steam chamber where contact with steam heats it to within just a few degrees of the saturation (flash) temperature. As feedwater temperature is elevated, the oxygen solubility is reduced by 97-98% and the dissolved oxygen dissociates. Oxygen and a small amount of steam are vented to the atmosphere. Deaeration is normally required when more that 15% make-up is required (i.e., less than 85% condensate recovery). Most units are rated at 0.005 $cm^3/1$ (7 ppb) of oxygen in feedwater. However, periodic feedwater testing should be done to ensure proper DA operation (2-4 times per year).

2.   **Deposition Control:** Just as for cooling towers, *blowdown* is the primary method of deposition control for boilers. The amount of blowdown is a function of the *quantity* and *quality* (hardness) of make-up water.

The blowdown requirement is determined by measuring boiler water electrical "conductance," a measure of the amount of conductive solids in the water (pure water has zero conductance). The recommended boiler water conductance is 3500 mega-mho/cm or less (where, electrically, a "mho" is the conductive equivalent to an "ohm" of resistance).

Two methods are used for boiler blowdown:

a.   Manual "bottom" blowdown consists of opening blowdown valves in accordance with an operating schedule dictated by periodic boiler water testing. This method removes both dissolved solids and sludge, but to be effective, Frequent short blowdown periods are required. The boiler operator must monitor boiler water level during blowdown to prevent boiler operating problems.

b.   Automatic "top" blowdown can be intermittent or continuous, as determined by conductance monitoring and the water treatment control system. With this method, blowdown is taken from the highest water level, where dissolved solids concentrations tend to be higher, resulting in a more efficient process. Heat recovery from the wasted boiler water is possible and is recommended. Note that with automatic blowdown, some manual blowdown may still be required for sludge removal.

The amount of boiler blowdown required is computed as follows:

$$\text{Required Blowdown (\%)} = 100 \times (A/B)$$

where

A   = actual dissolved solids concentration in the feedwater (mix of make-up and condensate)

B   = desired dissolved solids concentration in the boiler water, recommended as follows:

Boiler Operating Pressure (psig)	Total Dissolved Solids, TSD (ppm)	Total Alkalinity (ppm)
0-50	2500	500
51-300	3500	700
301-45	3000	600

Additional deposition control can be provided with chemical treatment by phosphate or phosphate/polymer, chelant, or chelant/polymer additives if blowdown fails to provide the required levels of TDS and/or alkalinity in the boiler water.

3. **Corrosion Control**: Corrosion in boiler systems occurs as the result of up to three mechanisms

a. Galvanic corrosion occurs when a metal is coupled with another metal of different electrical potential (valence). This condition induces an electrical current and loss of electrons at the cation. Galvanic corrosion can be caused by metallic scale deposits, surface pitting or scratches, etc. exposing different materials within the metal. This process is accelerated when two different metals are used, such as steel and copper.

b. Caustic corrosion occurs in the presence of a concentration of caustic (such as NaOH) due to steam blanketing that allows salts to concentrate on surfaces and/or localized boiling caused by porous deposits on tube surfaces. The caustic dissolves the protective magnetite layer, causing a loss of the underlying metal.

c. Acidic corrosion is caused by low feedwater pH and results in metal "thinning" (general corrosion) and/or local corrosion at bolts and other stress points

Corrosion in boiler systems is generally controlled by a combination of pH control (8.5-12.7 maximum range) and oxygen control (5-7 ppb) via deaeration and/or oxygen scavenging chemicals.

4. **Condensate System Corrosion**: Corrosion in condensate piping occurs due to two conditions

a. Oxygen "pitting," which results in localized loss of steel, occurs due to contact with air in atmospheric pressure pumped return systems. This problem can be treated by injection of an oxygen scavenger chemical into the condensate, but is a difficult and expensive method of treatment.

b. Acid corrosion occurs when carbon dioxide in air reacts with water to form carbonic acid that attacks steel. This corrosion is enhanced by decomposition of feedwater alkalinity that produces carbon dioxide. This problem can be treated by adding amines to feedwater to neutralize acids. Amines are introduced into the boiler water and "carry-over" into the condensate system with the steam.

5. **Boiler Fireside Problems**: Deposition and corrosion can occur on the fireside of boilers. Deposition occurs as deposits of fuel ash components on surfaces and may require treatment by fuel additives to dilute deposits. Corrosive fuel ash components (such as sodium sulfate, sodium vanadyl vanadate, etc.) may deposit on the boiler surfaces. These deposits have low melting points and, when in a liquid state, attack metal surfaces. Treatment additives to boiler fuels can raise melting points of deposits (but make deposits harder to remove). The best way to avoid this problem is to burn "clean" fuels (natural gas, light oils, low-sulfur/low ash heavy oils).

4.3.22.0—Bearing Lubrication

Lubrication of bearings on rotating equipment shafts (fans, pumps, etc.) must be done in strict accordance with the manufacturer's recommendations for type of lubricant and intervals between re-lubrication.

When no manufacturer's data are available, the re-lubrication interval, at least initially, can be estimated on the basis of the following:

Re-Lubrication Interval (Operating Hours) = $[14,000,000/(\text{RPM} \times \sqrt{D})] - (4 \times D)$

where

RPM = shaft rotational speed, revolutions per minute
D= shaft diameter at bearing, mm (where 1" = 25.4 mm)

This relationship is valid for tapered or spherical roller bearings. For cylindrical or needle bearings, multiply the result by 5. For ball bearings, multiply by 10.

This relationship is valid for bearing operating temperatures of 160°F or less. If bearing operating temperatures exceed 160°F, multiply the results by the following factor:

Bearing Operating Temperature (°F)	Multiplier Factor
≤160	1.0
170	0.75
180	0.5
190	0.375
200	0.25

Note that bearing environment and position may require more frequent re-lubrication—dirty, dusty environments, vertical shafts, etc. will shorten re-lubrication periods.

Ultimately, the proper re-lubrication period must be determined by "trial and error." Using manufacturer's data or the formula above as a starting point, visually examine purged lubricant at the end of lubrication interval. If the lubricant is clean, lengthen the period between re-lubrication. If it is dirty or scorched, shorten the interval.

Over-greasing of bearing assemblies will result in seal failure. Big globs of grease on the floor beneath the bearing is a sure sign this had happened. In this case, bearing replacement is required (and care must be taken in the future to avoid failure from recurring).

To prevent over-lubrication, there are two ways to determine the correct re-lubrication quantity. The first, and preferred, method is to contact the bearing manufacturer and request a recommendation based upon the specific application. Failing that, the second option is to calculate the correct re-lubrication quantity using the following equation

$$\text{Grease Quantity (oz)} = (D \times B \times Q)/28.35$$

Where

    D= Bearing outer diameter (mm)

    B= Bearing width (mm)

    Q= Re-lubrication interval factor (0.002 for a weekly re-lubrication interval, 0.003 for a monthly interval, 0.004 a yearly interval)

FIRE PROTECTION

4.4.1.0—Sprinkler Systems

All fire sprinkler systems must be inspected, tested, and maintained in accordance with NFPA Standard 25, *Standard for the Inspection, Testing, and Maintenance of Water-Based Fire Protection Systems.*

[Note: In many states, the personnel performing these functions are required to be licensed by the state, as follows:

Personnel performing **maintenance** *on fire sprinkler systems must hold a valid Limited Fire Sprinkler System Maintenance Technician license.*

Personnel performing **testing and inspection** *of fire sprinkler systems required by NFPA 25 shall hold a valid Limited Fire Sprinkler System Inspection Technician license.]*

4.4.3.0—Portable Extinguishers

Portable fire extinguishers shall be inspected and tested, as follows:

Monthly inspection:
1. Inspect when initially placed in service and thereafter at monthly intervals. Fire Extinguishers shall be inspected, manually or by electronic means, at more frequent intervals when circumstances require.

2. Inspection of fire extinguishers shall include a check of the following items:
    a. Location in designated place.
    b. No obstruction to access or visibility.
    c. Operating instructions on nameplate legible and facing outward.
    d. Safety seals and tamper indicators not broken or missing.
    e. Fullness determined by weighing or "hefting."
    f. Examination for obvious physical damage, corrosion, leakage, or clogged nozzle.
    g. Pressure gauge reading or indicator in the operable range or position.
    h. Condition of tires, wheels, carriage, hose, and nozzle checked (for wheeled units).

3.   Keep records of fire extinguishers inspected, including those found
      to require corrective action. The date the inspection was performed
      and the initials of the person performing the inspection shall be re-
      corded.

Annual Maintenance

1.   Stored-pressure types containing a loaded system agent shall be
      disassembled on an annual basis and subjected to complete mainte-
      nance.

2.   A conductivity test shall be conducted annually on all carbon diox-
      ide hose assemblies. Hose assemblies found to be nonconductive
      shall be replaced.

3.   Pressure regulators provided with wheeled-type fire extinguishers
      shall be tested for outlet static pressure and flow rate in accordance
      with manufacturer's instructions.

4.   At the time of the maintenance, the tamper seal of rechargeable fire
      extinguishers shall be removed by operating the pull pin or locking
      device. After the applicable maintenance procedures are completed,
      a new tamper seal shall be installed.

Six-Year Maintenance:

1.   Stored-pressure fire extinguishers that require a 12-year hydrostatic
      test shall be emptied and subjected to the applicable maintenance
      procedures. The removal of the agent shall only be done using a
      listed halon closed recovery system. When the applicable mainte-
      nance procedures are performed during a periodic recharging or hy-
      drostatic testing, the 6-year requirement shall begin from that date.

2.   Non-rechargeable fire extinguishers shall not be hydrostatic tested
      but shall be removed from service at a maximum interval of 12 years
      from the date of manufacture.

3.   Each fire extinguisher shall have a tag or label securely attached that
      indicates the month and year the maintenance was performed and
      that identifies the person performing the search.

Recharging

1. All rechargeable-type fire extinguishers shall be recharged after any use or as indicated by inspection or when performing maintenance.

2. After recharging, a leak test shall be performed on stored-pressure and self-expelling types of fire extinguishers.

3. Every 12 months, pump tank water and pump tank calcium chloride-based antifreeze types of fire extinguishers shall be recharged with new chemicals or water, as applicable.

4. The agent stored-pressure wetting agent fire extinguishers shall be replaced annually. Only the agent specified on the nameplate shall be used for recharging.

5. The premixed agent in liquid charge-type AFFF (aqueous film-forming foam) and FFFP (film-forming fluoroprotein foam) fire extinguishers shall be replaced at least once every three years. The agent in solid charge-type AFFF fire extinguishers shall be replaced once every 5 years.

6. Only those recharge agents specified on the nameplate or agents proven to have equal chemical composition, physical characteristics, and fire extinguishing capabilities shall be used. Agents listed specifically for use with that fire extinguisher shall be considered to meet these requirements.

7. Each fire extinguisher shall have a tag or label securely attached that indicates the month and year recharging was performed and that identifies the person performing the service.

Hydrostatic Testing

At intervals not exceeding those specified in the following table, fire extinguishers shall be hydrostatically retested.

The hydrostatic retest shall be conducted within the calendar year of the specified test interval. In no case shall an extinguisher be recharged if it is beyond its specified hydrostatic retest date. Hydrostatic testing shall be performed by persons trained in pressure testing procedures and safeguards who have suitable testing equipment, facilities, and appropriate servicing manual(s) available.

Extinguisher Type	Test Interval (Years)
Stored-pressure water, loaded system, and/or antifreeze Wetting agent Aqueous film-forming agent Film-forming fluoroprotein foam Dry chemical with stainless steel shell Carbon dioxide	5
Wet chemical Dry chemical, stored-pressure, with mild steel, brazed 　　brass, or aluminum shell Dry chemical, cartridge- or cylinder-operated, with mild 　　steel shell Halogenated agents Dry powder, stored-pressure, cartridge- or cylinder- 　　operated, with mild steel shell	12

1.　A hydrostatic test shall always include both an internal and external visual examination of the cylinder.

2.　Hydrostatic testing shall be conducted using water or some other noncompressible fluid as the test medium. Air or other gases shall not be used as the sole medium for pressure testing. All air shall be vented prior to hydrostatic testing to prevent violent and dangerous failure of the cylinder.

3.　Where a fire extinguisher cylinder or shell has one or more of the following conditions, it shall not be hydrostatically tested, but shall be condemned or destroyed by the owner or at the owner's direction:
   a.　Where repairs by soldering, welding, brazing, or use of patching compounds exist.
   b.　Where the cylinder threads are worn, corroded, broken, cracked, or nicked.
   c.　Where there is corrosion that has caused pitting, including pitting under a removable nameplate or name band assembly.
   d.　Where the fire extinguisher has been burned in a fire.
   e.　Where a calcium chloride type of extinguisher agent was used in a stainless steel fire extinguisher.

f. Where the shell is of copper or brass construction joined by soft solder or rivets.

g. Where the depth of a dent exceeds 1/10 of the greatest dimension of the dent if not in a weld, or exceeds ¼ in. (0.6 cm) if the dent includes a weld.

h. Where any local or general corrosion, cuts, gouges, or dings have removed more than 10 percent of the minimum cylinder wall thickness.

i. Where a fire extinguisher has been used for any purpose other than that of a fire extinguisher.

4. When a fire extinguisher cylinder, shell, or cartridge fails a hydrostatic pressure test, or fails to pass a visual examination, it shall be condemned or destroyed by the owner's agent. A condemned cylinder shall not be repaired. No person shall remove or obliterate the "CONDEMNED" marking.

5. Nitrogen cylinders, argon cylinders, carbon dioxide cylinders, or cartridges used for inert gas storage that are used as an expellant for wheeled fire extinguishers and carbon dioxide extinguishers shall be hydrostatically tested every 5 years.

6. A hydrostatic test shall be performed on fire extinguisher hose assemblies equipped with a shutoff nozzle at the end of the hose. The test interval shall be the same as specified for the fire extinguisher on which the hose is installed.

7. The pressure in a hydrostatic test of a cylinder shall be maintained for a minimum of 30 seconds, but for no less time than is required for complete expansion of the cylinder and to complete the visual examination of the cylinder. All valves, internal parts, and hose assemblies shall be removed and the fire extinguisher emptied before testing.

8. Recording of hydrostatic tests
   a. Cylinders or cartridges that pass the hydrostatic test shall be stamped with the retester's identification number and month and year of the retest. [It is important that the stamping be placed only on the shoulder, top head, neck, or footring (where provided) of cylinder.]

    b. Hose assemblies that pass a hydrostatic test do not require recording, labeling, or making.

    c. Fire extinguisher cylinders of the low-pressure type that pass a pressure hydrostatic test shall have the information recorded on a suitable metallic label or equally durable material with a minimum size of 2" x 3-½." The label shall be affixed by means of a heatless process. These labels shall be of the type that self-destructs when removal from a fire extinguisher shell is attempted.

4.4.3.1—Kitchen Hood Fire Suppression

Kitchen hood fire suppression systems must be inspected and maintained as follows:

Monthly

1. Check to make certain there is no corrosion to any of the detection system components (certain high alkaline cleaners could cause corrosion and their use must be eliminated).

2. Ensure that metal fusible links are replaced at least annually. Deterioration of these links could cause the system to be actuated or to malfunction in case of a fire.

3. Make certain the releasing unit has not been tampered with, and that visual inspection seals are not broken or missing.

4. Check system for loose pipes and missing or grease covered nozzle caps. Make certain nozzle caps are in place over the ends of each nozzle. Temporarily remove cap, check to make certain it is not brittle, and snap back on nozzle.

5. Check each metal blow-off cap and make certain the cap can be turned freely on the nozzle.

6. Check the visual indicator on the releasing unit to make certain the system is cocked.

7. Check that the manual pull station is not obstructed, has not been tampered with, and is ready for operation.

8.   Initial inspection tag. [Tags must be replaced when all lines are used or when tags are lost or removed.]

Every 6 months, preferably in August and February:
1.   Systems must be inspected and serviced by a manufacturer-authorized contractor.

Annually
1.   Replace fusible links on fire suppression system heads.

ELECTRICAL

WARNING! A fully documented "lock out/tag out" program is required for all electrical maintenance.

4.5.1.0—High-voltage Systems

High-voltage primary service and distribution (>600 volts) systems and devices shall be inspected and serviced annually as follows:

Distribution transformers (outdoor, oil-filled):
1.   Inspect solid electrical insulation for discoloration and degradation.

2.   Perform insulating oil test
    a.   Perform dielectric strength test in accordance with ASTM D 877.
    b.   Perform acidity test in accordance with ASTM D 1534.
    c.   Perform color test in accordance with ASTM D 1524.

3.   Verify cooling systems
    a.   Inspect forced cooling system equipment for damage, etc. Repair or replace as necessary.
    b.   Operate system by simulating high temperature at controlling devices.

4.   Verify transformer relay protection.

5.   Verify transformer alarms.

Circuit breakers and switchgear
1.   Inspect switchgear
    a.   Inspect, investigate, and solve condition(s) causing carbon tracks.

b. Inspect barriers and shutters for physical damage. Prove shutter operation, if possible.

2. Test switchgear phase bus insulation
    a. Perform insulation resistance test on each phase-to-phase and phase-to-ground using a megohmmeter.
    b. Perform dielectric absorption test on each phase using a megohmmeter.
    c. Perform DC over-potential test on each phase-to-phase and phase-to-ground.
    d. Perform power factor test on each phase.

3. Service circuit breakers
    a. Inspect draw-out contacts for abnormal wear, tension, or discoloration. Correct as required.
    b. Inspect breaker current-carrying components for overheating discoloration. Replace as required.
    c. Inspect, operate, adjust, and lubricate mechanical linkages. Replace any components at are damage, rusted, etc.
    d. Verify the opening and closing sequence of arcing, intermediate, and main contacts on air circuit breakers.
    e. Verify interlocks preventing a closed breaker from being withdrawn from or connected to the switchgear bus.
    f. Inspect and dress current carrying contacts on air circuit breakers in accordance with the manufacturer's requirements.
    g. Inspect the contact wear indicator on vacuum circuit breakers. Replace the vacuum bottle as required.

4. Test circuit breakers
    a. Perform test operations to prove correct actuation of breakers' trip and close components, including spring charging motors, trip solenoids, indicating targets, etc.
    b. Perform contact resistance test.
    c. Perform insulation resistance test on each phase-to-phase and phase-to-ground using a megohmmeter.
    d. Perform DC over-potential test on each phase-to-phase and phase-to-ground.
    e. Perform voltage test across each open contact of vacuum circuit breaker to verify vacuum condition of supply bottle.

   f. Prove circuit breaker operation by actuation of each associated
       protective relay.

   g. Prove circuit breaker operation by actuation of each breaker
       manual control switch.

5. Verify switchgear alarms.

6. Test air switch
   a. Perform DC over-potential test on each pole-to-pole and each
       pole-to-ground.
   b. Perform contact resistance test across each switch and fuse
       holder.

7. For bolted bus, torque test connections. Clean bus insulators and
   perform resistance test for ground.

Power Cables
1. Inspect cable installations for the following:
   a. Accessible portions of cables, especially splices and termina-
       tions, must be visually inspected for insulation damage, carbon
       tracking, discoloration, signs of corona, etc. Replace any cables
       with problems.
   b. Inspect cable shield grounding equipment such as conductors
       and connections. Repair as required.

2. After 10 years in service, and every 3 years thereafter, test and evaluate insulation of primary shielded power cables by using Type 2
   (Diagnostic) Field Tests in accordance with IEEE Standard 400.

4.5.1.1—Low-voltage Systems
   Low voltage (≤600 volts) systems and devices shall be inspected and
serviced as follows:

Facility transformers (indoor)
1. Inspect, clean, and service *dry type* transformers as follows:

Monthly
   a. Check operating temperature (150°C max. for transformers rat-
       ed for 80°C rise, 220°C max. for transformers rated for 150°C
       rise).

    b. Check and clean ventilation louvers.

    c. Test that loading is within transformer rating.

Every 5 years

    a. Inspect base or support insulators.

    b. Tighten all bolted connections.

    c. Inspect coil windings for damage or loose iron; clean as needed.

    d. Examine primary, secondary, and ground connections.

    e. Examine insulation for signs of over-heating.

    f. Clean, check porcelain insulators.

    g. Test high-to-low to ground resistance.

    h. Inspect tap connections and tap changer.

    i. Clean enclosure via vacuum or blower.

2. Inspect and clean *oil-filled* transformers every 3 years, as follows:

    a. Sample and test transformer oil and test insulation.

    b. Examine primary, secondary, and ground connections.

    c. Inspect handhole cover gaskets for seal.

    d. Inspect seals on tap changer compartment.

    e. Inspect radiators for damage, oil leaks, rust, dirt. Clean as required.

    f. Inspect, test sudden pressure relay.

    g. Check, test temperature gauge.

    h. Inspect switch and terminal chambers.

    i. Inspect condition and confirm operation of tap changer and position indicator.

Panels, circuit breakers, and switchgear. Perform the following inspections, tests, and maintenance every 6 months:

1. Inspect and test

    a. Mechanical Checks

        1) Examine and tighten connections, incoming, outgoing cables, phase, neutral and ground.

        2) Fuses, tightness of fuse clips.

        3) Mechanical operation of all circuit breaker or switch operating mechanisms.

        4) Interior firmly bolted in cabinet and aligned.

        5) Broken or missing parts from breakers or switch devices.

      6) Missing hardware from the internal assembly, cabinet or outer trim. Outer trim accurately aligned and securely fastened.

  b. Inspection Checks

      1) Discoloration at fuse clips and device terminals.

      2) Charring of molded insulating parts, including molded breaker cases, fuse holders, wire insulation.

      3) Evidence of moisture, contamination, dust, hardware, debris, etc. Cracking, peeling, brittleness of cable, wire insulation.

      4) Clearances to grounded metal and between phases of installed devices and cable/wire.

      5) Abrasion of cable and wire insulation. Nicking of wire or cable strands at terminals.

      6) Overcrowding in wiring gutters.

      7) Adequate sealing of conduits and cabinet openings.

  c. Electrical Checks

      1) Fuses, check, verify type, voltage, current and interrupting ratings. Resistance test, phase to phase to ground, insulation.

      2) On service entrance type, neutral-to-ground bonding conductor in place and securely connected.

      3) Equipment grounding terminal bar securely connected to cabinet or panelboard frame.

      4) Proper connection of ground fault devices, pushbutton test of ground fault breakers.

2. Perform infrared thermography inspection every 3 years on outdoor panels and every 5 years on indoor panels.

Motor control centers (MCC) and motor starters. Perform the following inspections, tests, and maintenance every 6 months

1. Inspect MCC and starters for the following:

  a. Inspect main contacts and auxiliary contacts on contactors for evidence of arcing, fusing, etc. Replace contactors as required.

  b. Inspect pushbuttons, indicating lights, selector switches, etc. and test for proper operation. Replace as required.

2. Test phase bus insulation.

3. Service MCC
    a. Inspect draw-out contacts for abnormal wear, tension, or discoloration. Correct as required.
    b. Inspect unit's current-carrying components for overheating discoloration. Replace as required.
    c. Inspect, operate, adjust, and lubricate mechanical linkages. Replace any components at are damage, rusted, etc.
    d. Verify mechanical interlocks.
    e. Inspect and dress current carrying contacts on switches and contactors in accordance with the manufacturer's requirements.

4. Test motor starter
    a. Manually operate switches and circuit breakers to verify correct operation.
    b. Operate starter using all manual and automatic control devices to ensure correct operation.
    c. Verify correct interlocking action with other associated equipment.
    d. Verify correct indicating light operation during each test.

5. Verify equipment alarms

Disconnects and safety switches. Perform the following inspections, tests, and maintenance every 6 months

1. Inspect safety switches for the following:
    a. Inspect, operate, adjust, and lubricate mechanical linkages. Replace any components at are damage, rusted, etc.
    b. Verify operation of mechanical interlocks.
    c. Inspect and dress current carrying contacts on switches and contactors in accordance with the manufacturer's requirements.

2. Test safety switches
    a. Perform insulation resistance test on each phase-to-phase and phase-to-ground using a megohmmeter.
    b. Perform contact resistance test.

Ground fault circuit interrupter (GFCI) receptacles and circuit breakers
1. Test monthly for proper operation. Repair or replace as required.

Electrical equipment rooms and closets

1.   Inspect monthly and maintain required clearance in front of electrical panels.

Distribution Feeders
1.   Inspect accessible portions of cables, especially splices and terminations, must be visually inspected for insulation damage, carbon tracking, discoloration, signs of corona, etc. Replace any cables with problems.

2.   Perform cable insulation test every 3 years.

3.   Perform busway insulation test every 5 years.

Receptacle loads
1.   Inspect cords and appliances annually for proper use, damage, over-loading, etc. Replace damaged or unsafe cords. *Remove over-loaded cords...*create work order to add receptacles and/or receptacle circuits as required.

4.5.2.1—Lighting Systems
     Inspect, test, and maintain lighting systems as follows:
1.   Monthly, perform visual inspection
     a.   Inspect fixtures to identify inoperable or faulty lamps and ballasts. Replace as required.
     b.   Inspect fixtures to identify excessive dirt. Clean as needed (see below).
     c.   Test lighting controls for proper operation. Adjust or replace as required.
     d.   Replace yellowed, stained, or broken lenses or louvers.
     e.   For gym lights, replace missing or broken light guards.

2.   Annually, clean light fixtures, including lamps, as follows:
     a.   Turn off lights and allow lamps and fixtures to cool.
     b.   Use very mild soaps and cleaners, followed by a clean rinse. Avoid strong alkaline cleaners or abrasive cleaners. Silver films require the mildest 0.5% solution and a soft, damp cloth.
     c.   Glass cleaners may be used on porcelain or glass, but require an additional clean rinse.

d. To avoid static charge on plastics, use anti-static cleaning compounds. Do not wipe plastics dry after rinse, air dry with vacuum cleaner or fan.

3. Annually, inspect, test, and calibrate light control devices in accordance with the following:

Control Device	Calibration Setting(s)
Occupancy sensor, ceiling-mounted	Time delay: 15 minutes (30 minutes in large spaces such as classrooms) Sensitivity: Medium high
Occupancy sensor, wall-mounted	Manual: On Auto: Off Time delay: 15 minutes (30 minutes in large spaces such as classrooms) Sensitivity: Medium
Manual dimmer (Incandescent only)	High end trim set at 95%
Automatic timers, timer switches	On/Off times based on occupancy set for each day of the week Holiday and weekend schedules

4. Group relamp all fixtures at 70% of average rated lamp life in accordance with the following schedule

Lamp Type	Avg. Annual Avg. Rated Lamp Life (Hours)	Utilization (Hours)	Group Relamping Frequency
Fluorescent	12,000	2500	3.3 years
Metal Halide	15,000	3000	3.5 years
Mercury Vapor	24,000	3000	5.6 years
HP Sodium	20,000	3000	4.7 years
Incandescent	750	2500	4 months

4.5.3.0—Public Address System

Inspect, test, and maintain public address system in accordance
with manufacturer's written instructions.

4.5.3.1—Intercom System

Inspect, test, and maintain intercommunication/paging system in accordance with manufacturer's written instructions.

4.5.3.2—Call System

Inspect, test, and maintain call system in accordance with manufacturer's written instructions.

4.5.3.3—Clock System

Inspect, test, and maintain clock system in accordance with manufacturer's written instructions.

4.5.3.4—Fire & Smoke Detection

Test and maintain fire and smoke detection and alarm systems as follows:

1. *Battery-powered single station smoke detectors*
    a. Monthly, test for proper operation. Replace battery or head if there is a malfunction.
    b. Annually, replace battery.
    c. Every 5 years, conduct calibration test for sensitivity.
    d. Every 10 years, replace detectors.

2. *Central fire alarm system*: Fire alarm system shall be maintained in accordance with the requirements of NFPA 72. Equipment and devices shall be tested in accordance with Table 10.4.3 of NFPA 72.

4.5.3.5—Security & Alarm System

Inspect, test, and maintain security and alarm system in accordance with manufacturer's written instructions.

4.5.3.6— Closed-circuit Systems

Inspect, test, and maintain closed circuit TV (CCTV) system in accordance with manufacturer's written instructions.

4.5.4.0—Grounding

Inspect electrical grounding connections, grounding rods, and wiring every 5 years. Tighten connections. Make sure that water piping systems used for grounding have no plastic piping elements.

4.5.4.1—Emergency Lighting

Emergency lighting systems require periodic inspection and maintenance, as follows:

1. Monthly, test emergency lighting system(s) for 30 seconds to ensure proper operation. Repair or replace any defective lamps, fixtures, or other devices.

2. Perform annual inspection, cleaning, and 90 minute operational test. Repair or replace any defective lamps, fixtures, or other devices.

4.5.4.2—Emergency Power

Emergency power systems require periodic inspection and maintenance, as follows:

Motor generators

1. Inspect motor generator sets for the following:
    a. Inspect to ensure that warning signs exist. If missing, replace.
    b. Inspect enclosure for damage, unauthorized openings, and corrosion. Repair and repaint a needed.
    c. Inspect air passages for blockages.
    d. Inspect, investigate, and solve conditions for unusual odors.
    e. Inspect electrical connections for tightness and degradation. Tighten and repair as required.
    f. Inspect electrical insulation for discoloration and degradation. Repair as required.
    g. Inspect equipment grounding components, conductors, and connections. Repair as required.
    h. Inspect locking devices. Repair as required.

2. Clean equipment.

3. Tighten all electrical connections.

4. Test motor and generator insulation
    a. Perform insulation resistance tests using a megohmmeter in accordance with IEEE 43 on the stator and rotor of motor, generator, and exciter.
    b. Perform dielectric absorption testing using a megohmmeter.

5.   Perform infrared test.

6.   Perform load test
   a.   Verify frequency and voltage output.
   b.   Verify instrumentation for correct indications.
   c.   Listen, investigate, and solve conditions for unusual noises.

7.   Inspect bearings
   a. Verify bearings are lubricated in accordance with manufacturer's instructions.
   b. Perform vibration tests.
   c. Check alignment and couplings.

8.   Measure and record neutral current.

9.   Verify system controls
   a.   Using calibrated test instruments, calibrate ammeters, voltmeters, etc.
   b. Verify continuity of metering selector switch contacts with ohmmeter.
   c. Run controller diagnostics (if provided).
   d. Simulate automatic and manual control sequences.
   e. Verify alarms.

Automatic transfer switches
1.   Inspect switches for the following:
   a. Inspect, operate, adjust, and lubricated mechanical linkages. Replace components as required.
   b. Verify operation of mechanical interlocks.
   c. Inspect and dress current carrying contacts in accordance with manufacturer's instructions.

2.   Test switches
   a.   Perform insulation resistance test on each phase-to-phase and phase-to-ground using a megohmmeter.
   b.   Perform contact resistance test.
   c.   Prove correct operation of the switch by manually initiating transfers in both directions.
   d.   Simulate automatic conditions to initiate automatic transfer in both directions.

e. Verify starting of generators (where applicable).

f. Verify correct indicating light operation.

3. Verify equipment alarms.

4.5.4.3—Uninterruptible Power Supply

Inspect, test, and service static uninterruptible power supply systems as follows:

Monthly

1. Perform visual checks of the following on batteries: Jar seals, terminal condition, battery plate discoloration, sediment, mossing, gassing, electrolyte level, vents, and flame arresters.

2. Inspect and test batteries: Measure specific gravity of electrolyte in pilot cell and correct for temperature.

3. Record battery bank charge voltage, type of charge, and charging current. Compare to last month's values. If there is a significant difference, investigate, determine problem(s), and make repairs.

4. Inspect battery rack
    a. Inspect for corrosion. Repair as required.
    b. Inspect for loose connections. Tighten or repair as required.
    c. Verify that battery rack is grounded.

5. Inspect ventilation filters and change or clean as needed.

6. Using a volt-ohm meter, verify correct operation of all switches.

7. Calibrate metering.

8. Test control cabinet
    a. Test indicating lamps and replace as required.
    b. Test annunciator's local visual and audible alarm.

9. Inspect battery charger for abnormalities.

10. Run equipment diagnostics routine and correct any abnormalities that are recorded.

Quarterly

1. Perform battery cell tests
   a. Measure specific gravity of each cell's electrolyte and correct with past readings. Compare with last tests. If there is a significant difference, investigate, determine problem(s), and make repairs.
   b. Record the voltage of each cell. Evaluated and compare with past readings.
   c. Record a sampling of intercell connector resistances.

2. Perform a complete load test in accordance with IEEE 450.

Annually

1. Inspect system for the following:
   a. Inspect to ensure that warning signs exist. If missing, replace.
   b. Inspect enclosure for damage, unauthorized openings, and corrosion. Repair and repaint a needed.
   c. Inspect, investigate, and solve conditions for unusual odors or noises.
   d. Inspect electrical connections for tightness and degradation. Tighten and repair as required.
   e. Inspect electrical insulation for discoloration and degradation. Repair as required.
   f. Inspect equipment grounding components, conductors, and connections. Repair as required.
   g. Inspect locking devices. Repair as required.

2. Tighten electrical connections.

3. Clean equipment.

4. Test phase bus insulation.

5. Perform infrared test.

6. Perform harmonic analysis.

7. Verify alarm setpoints and control limits
   a. Calibrate alarm setpoints such as temperature, over-voltage, over-current, etc.

    b.   Calibrate control limits for static switch operation, cooling fan operation, cooling water control, oscillator, load sharing controls, etc.

    c.   Verify correct action of any associated remote alarm.

8.   Measure and record neutral currents.

4.5.4.4—Rotary Uninterruptible Power

Inspect, test, and service rotary uninterruptible power supply systems as follows:

Quarterly

1.   Perform load test.

Annually

1.   Test induction coupler and synchronous machine insulation

    a.   Perform insulation resistance tests using a megohmmeter in accordance with IEEE 43 on the stator and rotor.

    b.   Perform dielectric absorption testing using a megohmmeter.

2.   Inspect bearings for induction coupler, synchronous machine, and free-wheeling coupler

    a.   Verify bearings are properly lubricated. Re-lubricate as required in accordance with manufacturer's instructions.

    b.   Perform vibration tests.

    c.   Check alignment.

Every 5 Years

1.   Replace free-wheeling coupler.

Appendix E

Equipment & Furnishings Preventative and Predictive Maintenance Procedures

Facility equipment and furnishings include permanently installed equipment, such as parking control devices, food service freezers and coolers, dock levelers, kitchen hoods, etc.

5.1.3.0—Automatic Gages
Every 60 days, inspect, test, and service automatic gates as follows:

1. Inspect gate arm. If damaged, replace.

2. Check belt tension. Belts should be tight to assist in controlling over-travel after operation of limit switches.

3. Adjust up and down limit switches in accordance with manufacturer's written instructions.

4. Lubricate gears and drive motors in accordance with manufacturer's written instructions.

5. Tighten all nuts, bolts, and screws in the entire gate unit.

6. Test gate operation by placing AUTO-MANUAL function switch in MANUAL position and testing in accordance with manufacturer's written instructions.

7. Every other inspection, clean and wax gate control housing or cabinet.

5.1.3.1—Dock Levelers & Locks
Every 3 months, inspect and repair dock levelers and locks:

1. Check all safety devices on the unit

2. Clean the upper portion of the plate, lower rollers and between the sides and curb angles.

3. Clean all debris from the pit or from the vicinity of floor mounted units in order to avoid interference with the rolling mechanism.

4. Inspect the push-in bar and push-out plate assemblies for damage.

5. Inspect the safety leg system and return spring operation.

6. Check for presence and proper setting of all snap rings and clips on axles and rollers.

7. Check rollers, pins and bushings for any signs of wear such as flat spots, missing fasteners or dislodged bearing material.

8. Check the springs for elongation.

9. Inspect all welds under the leveler for fatigue or failure, particularly the top plate understructure.

10. Check general condition of the dock leveler. Operate dock leveler after service, check for any abnormal noise or vibrations.

5.1.4.0—Built-in Kitchen Coolers
Weekly, inspect built-in kitchen coolers/freezers and service as required:

Reach-in refrigerators/freezers:
1. Check condition of door gaskets. Replace if there is any ice build-up around doors or gaskets are torn, flattened, or separated from door (loose). Adjust door hinges as required for tight shutting.

Walk-in coolers/freezers:
1. Check condition of door gaskets. Replace if there is any ice build-up around doors or if gaskets are torn, flattened, or separated from door (loose). Adjust door hinges as required for tight shutting. [Observe door use by kitchen staff. If doors are left open excessively, replace existing hinges with self-closing type.]

2. Check heater wires for proper function by feeling for warmth around door frames, perimeters of doors, and pressure relief ports. Ice buildup around doors indicates leakage and/or heater failure.

3. Check thermometers to ensure that temperature is being maintained. Every 6 months, test thermometers for accuracy and replace if error exceeds 5%.

4. Monthly, check refrigeration system:
    a. Read system pressures and, if necessary, add refrigerant. Inspect for refrigerant leaks.
    b. Check defrost cycle timeclocks to see that they are set and operating properly. Adjust length of defrosts on a seasonal basis.
    c. Check condenser fan(s) and fan motor(s) for proper operation. [Typically, motors are permanently lubricated. However, if re-lubrication is required, do so in accordance with Maintenance Procedure 4.3.20.0.] Clean condenser coils.
    d. Check evaporator fan(s) and fan motor(s) for proper operation. [Typically, motors are permanently lubricated. However, if re-lubrication is required, do so in accordance with Maintenance Procedure 4.3.20.0.] Clean evaporator coils.
    e. Check drain pan, drain lines, and drain heater for clogging, freezing, and proper operation. Blow out lines as needed.

5. Annually, inspect and tighten all electrical connections.

5.1.4.1—Kitchen Hoods
   Inspect and maintain kitchen hood systems as follows:

Routinely:
1. Ensure that cooking equipment is not operated while the hood fire extinguishing system or exhaust system is non-operational.

2.   Ensure that hood exhaust systems is operated whenever cooking equipment is turned on.

3.   All filters shall be in proper placement when exhaust system is in operation.

4.   Instructions for manually operating the fire extinguishing system must be posted conspicuously in the kitchen and must be periodically reviewed with employees by the management.

Weekly:
1.   Check filters. Mesh filters are no longer allowed by NFPA 96 because they do not drain well and clog up with grease quickly. Replace mesh filters with new UL listed grease filters.
2.   If filters are the proper type, clean by pressure washing where all grease and water will drain to the grease trap. [An alternative to pressure washing is to soak the filters in a water, detergent bath in a pot sink overnight, then rinsing thoroughly.]
3.   Check and service exhaust (and make-up) fan(s) in accordance with Maintenance Procedure 4.3.10.1.

Monthly:
     Inspect and maintain hood fire protection system in accordance with Maintenance Procedure 4.4.3.1.

Annually:
1. Clean hood and exhaust ductwork, as follows:
     a. All hot water high-pressure cleaning units used must put out water in excess of 180°F and 1500 psig.
     b. Before cleaning, inspect the exhaust system with the manager using the service report. Note any problems found such as bad wiring, grease on roof, fan noise/vibration, etc.
     c. Verify the fans are in good working condition.
     d. Cover the grease guard or other grease collection devices on the roof with plastic.
     e. Protect and cover all equipment in the kitchen service area with plastic.
     f. Bag or drape the hood with plastic to funnel all the water and grease into a container to avoid getting any on the equipment

or on the floor.

g. Apply degreasing chemical to the fan, all ductwork, filters, and the hood.

h. Clean the hood, filters, all vertical and horizontal ductwork and the fans.

i. Empty and clean the catch pans on the fans.

j. Wipe down the roof ladder and roof hatch, all equipment in the cleaning area and kitchen mop the floor.

k. Polish the inside and outside of the hood, and the backsplash (as applicable).

l. Make sure the fans and all kitchen equipment are working properly after cleaning.

m. Ensure that service sticker on the hood is replaced with new, properly dated sticker..

2. The cleaning outlined above should be done by a contractor with the required training and equipment.

3. Secure fire suppression system during cleaning. After cleaning, have fire suppression system serviced in accordance with Maintenance Procedure 4.4.3.1.

Appendix F

Special Construction Preventative and Predictive Maintenance Procedures

Special construction includes portable or temporary buildings, pre-engineered metal buildings, aquatic facilities, etc.

6.1.0.0—Pre-engineered Metal Buildings

Inspect and maintain all pre-engineered metal buildings in accordance with the applicable maintenance procedures of Sections 1-7.

6.1.0.1—Portable Classrooms/Temporary Buildings

Inspect and maintain all portable classrooms or other temporary buildings in accordance with the applicable maintenance procedures of Sections 1-5 of this chapter, *except reduce the time between inspections and maintenance elements by 50%.*

6.2.0.0—Interior Building Structure

Every 6 months, inspect interior building structure, all interior finishes, exposed HVAC elements, and equipment for corrosion. If the rusting is merely a surface accumulation or flaking, then the corrosion is *light*. If the rusting has penetrated the metal (indicated by a bubbling texture), but has not caused any structural damage, then the corrosion is *medium*. *If the rust has penetrated deep into the metal, the corrosion is heavy. Heavy corrosion generally results in some form of structural damage, through delamination, to the metal section, making repair difficult and expensive and requires the input of a Professional Engineer.*

1. Remove light and medium rust via manual abrasion (wire brush and aluminum oxide sandpaper) and mechanical abrasion (such as an electric drill with a wire brush or a rotary whip attachment).

Rust can also be removed by using a number of commercially prepared anticorrosive acid compounds. If chemicals are used, any chemical residue should be wiped off with damp cloths, then dried immediately with industrial blow-dryers. *Do not use running water to remove chemical residue.*

2.   Removing rust will remove most flaking paint as well. Remaining loose or flaking paint can be removed with a chemical paint remover or with a pneumatic needle scaler or gun. (Well-bonded paint may serve to protect the metal further from corrosion and need not be removed. The paint edges should be feathered by sanding to give a good surface for repainting.)

3.   Once metal has been cleaned of all corrosion, small holes and uneven areas resulting from rusting should be filled with a patching material of steel fibers and an epoxy binder and sanded smooth to eliminate pockets where water can accumulate.

4.   Bare metal should then be wiped with a cleaning solvent such as denatured alcohol, and dried immediately in preparation for the application of an anticorrosive primer (oil-alkyd based paint rich in zinc or zinc chromate), applied immediately after cleaning.

6.2.0.1—Oil Storage Tanks
Inspect and maintain oil storage tanks and appurtenances as follows:

Under ground storage tanks:
1.   Monthly, inspect to confirm that tank is not leaking. Monitor tank(s) for leaks by one or more of the following methods:
   a.   Daily inventory control: Measure deliveries, withdrawals, and oil amount in tank by manual "sticking" (tanks ≤ 2,000 gallons only) or level gauging (a level measuring system capable of measuring oil depth to within 1/8")
   b.   Monthly manual "sticking" (tanks ≤ 2,000 gallons only). Measurements are required at the beginning and end of a 36 hour period during which no liquid is add to or removed from the tank.
   c.   Monthly tank tightness test.
   d.   Continuous automatic tank gauging with a system capable of

detecting leaks of 0.2 gph leak rate in conjunction with daily inventory control.

e.   Groundwater monitoring.

f.   For double wall tanks, interstitial space monitoring.

2.   Every 60 days, as applicable, inspect impressed current cathodic protection system to ensure the equipment is functioning properly.

3.   Every 3 years, as applicable, inspect and test sacrificial cathodic protection system in accordance with National Association of Corrosion Engineers Standard RP-02-85.

4.   Annually, monitor piping for leaks by a line tightness test.

Above ground storage tanks:
Weekly:
1.   Inspect tank and exposed piping for leaks. Check containment area for liquid in the containment area or liquid at the tank seams, connections, or piping joints. If any leaks are observed, stop use of tank, drain it, and correct leaks.

2.   Inspect tank and exposed piping for corrosion (rust). If the rusting is merely a surface accumulation or flaking, then the corrosion is *light*. If the rusting has penetrated the metal (indicated by a bubbling texture), but has not caused any structural damage, then the corrosion is *medium*. *If the rust has penetrated deep into the metal, the corrosion is heavy. Heavy corrosion generally results in some form of structural damage, through delamination, to the metal section, making repair difficult and expensive and the tank should be replaced.*

a.   Remove light and medium rust via manual abrasion (wire brush and aluminum oxide sandpaper) and mechanical abrasion (such as an electric drill with a wire brush or a rotary whip attachment). Rust can also be removed by using a number of commercially prepared anticorrosive acid compounds. If chemicals are used, any chemical residue should be wiped off with damp cloths, then dried immediately with Industrial blow-dryers. *Do not use running water to remove chemical residue.*

b.   Removing rust will remove most flaking paint as well. Remaining loose or flaking paint can be removed with a chemical paint remover or with a pneumatic needle scaler or gun. (Well-bonded paint may serve to protect the metal further from corrosion and need not be removed. The paint edges should be feathered by sanding to give a good surface for repainting.)

c.   Once metal has been cleaned of all corrosion, small holes and uneven areas resulting from rusting should be filled with a patching material of steel fibers and an epoxy binder and sanded smooth to eliminate pockets where water can accumulate.

d.   Bare metal should then be wiped with a cleaning solvent such as denatured alcohol, and dried immediately in preparation for the application of an anticorrosive primer (oil-alkyd based paint rich in zinc or zinc chromate), applied immediately after cleaning.

3.   Inspect containment area:
a.   If retained storm water is present *and* there is no evidence of oil leaks, unlock and open drain valves. Once water is drained, close and *lock* drain valves.
b.   It there is leaves, trash, clutter, or waste in the containment area, clean the area.

Monthly:
1.   Inspect piping, fittings, pumps, valves, gauges and level monitoring devices, meters, etc. for condition, operation, leaks, and corrosion. Repair or replace faulty or corroded devices. Repair, any leaks that are found.

2.   Inspect containment area for cracks or gaps. Repair any found.

3.   Complete monthly inspection report required for State or Federal record-keeping regulations.

Appendix G

Sitework Preventative and Predictive Maintenance Procedures

Sitework includes site improvements (roads, parking lots, walkways, landscaping, etc.) and civil, mechanical, and electrical utilities.

SITE IMPROVEMENTS

7.1.1.0—Gravel or Shellrock Paving

Every 3 months, inspect and maintain gravel or shellrock (marl) paving as follows. If possible, conduct maintenance operations when moisture is present and surface is wet and soft.

WHAT TO LOOK FOR	PROBABLE CAUSE	REMEDIAL ACTIONS
Deteriorated crown and surface drainage patterns (secondary ditches)	Inadequate crown and resulting poor drainage	Re-grade with 4-6% slope from the road centerline to the shoulder.
Need to add and/or distribute gravel or shellrock	Road wear, erosion, etc.	Prepare road surface by scarifying it lightly. Have trucks deliver new material and place it as uniformly as possible. With grader, distribute material evenly, careful to maintain crown. Allow roadway traffic to compact new material.

Corrugation/washboarding	Road traffic, poor crowning and material distribution, drainage across road.	Cut out the road and reconstruct, being carefully to remix and compact fine and course materials.
Potholes	Poor drainage	Locate and fix drainage problem first. Then, with grader, cut out the potholes, place additional material as needed, and grade and compact surface, careful to maintain crown.
Rutting	Road traffic, poor drainage	With grader, cut out the rutted roadway, place additional material as needed, and grade and compact surface, careful to maintain crown.

7.1.1.1—Concrete Paving

Annually, inspect and maintain concrete paving as follows:

WHAT TO LOOK FOR	PROBABLE CAUSE	REMEDIAL ACTIONS
Joint deterioration/ spalling	Caused when water and debris fill the joint space, putting pressure on the concrete.	Joint repair or full-depth repair.
Cracks	Stress cracks include trans – verse cracks that run across the pavement, perpendicular to the shoulder and longitudinal cracks that run parallel to the shoulder. D cracking occurs at slab corners where longitu – dinal and transverse joints intersect. The failure is due to poor quality aggregate in the original concrete mixture. due to poor quality aggregate in the original concrete mixture.	Joint/crack filling or full-depth repair.

	Map cracking is a pattern of inter-connected random cracks that indicate failure in the sub-base.	
Blowups	Upward movement of the pavement surface at transverse joints or cracks, generally during hot weather. They are generally caused by a buildup of pressure in the pavement, which causes the panels on each side of a joint to rise.	Temporary repair or full-depth repair.
Scaling	Deterioration of the upper concrete slab surface, normally 3/8" to 1/2" deep, and may occur anywhere on the pave ment. It may be caused by deicing chemicals or by inadequate application of curing compound.	Surface patching or temporary repair.
Settlement	Slabs sometimes settle, parti-cularly bridge approach panels. A slab with tilted or uneven panels may indicate that subbase materials have migrated from beneath the slab.	Temporary repair or mud jacking.
Faulting	A difference in elevation across a joint or crack caused by the settlement of one or both of the slabs or by rocking of the slabs as traffic moves across the joint or crack.	Temporary repair or mud jacking.
Pumping	Seeping or ejection of water and sub-base from beneath the pavement through cracks.	Joint repair.
Corner breaks	A corner portion of the slab may separate from the slab along a crack that intersects the adjacent transverse and longi tudinal joints at an approximately 45-degree angle.	Joint repair or full-depth repair.
Pavement failure	Pavement is considered to have failed when the deterioration of the pavement becomes so severe that the only option for repair is to remove and replace the slab.	Full-depth repair.
Pavement marking	Wear and deterioration	Apply new markings in accordance with Maintenance Procedure 7.1.1.4.

Joint/crack filling:

1.   Rout or saw-cut joints to remove existing sealant and other materials
     and to provide clean, uniform surfaces for filler to adhere to and a
     reservoir for sealant. Rout joints to a width of to ½ inch and a depth
     of ¾ to 1 inch.

2.   If feasible, sandblast the joints.

3.   Use an air compressor and an air wand to clean joints of dirt, dust, and
     remnants from sawing/routing and sandblasting. Contamination in
     a joint will cause poor sealant bonding.

4.   If using backer rods in joints, place the rod to the proper depth to
     ensure the correct shape of the sealant reservoir. On road surfaces
     where grinding is planned at a later date, the backer rod and sealant
     should be installed so that the sealant is approximately ¼ inch below
     the road surface after grinding is complete.

5.   Apply sealant according to specifications and the manufacturer's
     recommendations. The joint will be filled with a concave bead. The
     shape factor generally ranges from 1:2 to 2:1, depending mainly
     on the elasticity of the sealant material. Be guided by the type of
     material and its specifications on determining the proper shape
     factor.

Joint repair:

1.   Saw cut, break out, and remove loose material, leaving the faces of
     the removal vertical. A cutting torch or saw may be necessary for
     cutting pavement reinforcement. Normally the steel network is not
     reestablished.

2.   Clean the hole with compressed air to remove moisture and debris.

3.   Fill the hole with concrete mix, normally delivered by a ready-mix
     operation.

4.   Consolidate the mix with a vibrator.

5.   Screed and finish the surface, but do not add free water. (Adding

free water to the surface dilutes the cement paste, and the surface is likely to scale off in the near future.)

6. Cure the concrete by covering with plastic, wet burlap, or a liquid curing compound. The burlap should be kept wet until the initial concrete strength is developed.

Full-depth repair:

Apply a full-depth repair where the depth of the deterioration is greater than 25 percent of the total pavement thickness or covers a large area.

1. Mark the area to be patched two to three inches outside the damaged area.

2. Saw cut and remove full depth of concrete slab without damaging adjacent concrete.

3. Remove any unsound base or sub-base. If a pre-cast slab is to be used, the base or sub-base needs to be restored and compacted. If a serious drainage problem exists, it should be corrected as with transverse, lateral sub-drain, etc.

4. Other than pre-cast, place a form for reestablishing shoulder edge.

5. Sandblast exposed concrete and clean area with compressed air.

6. Use coated dowel bars and deformed rebars for load transfer in all full-depth repairs.

7. Place low-slump concrete with adequate mechanical vibratory screeds.

8. Texture and cure the concrete.

Surface patching:

Apply a surface patch where the depth of deterioration is no more than 25 percent of the total pavement thickness:

1. Mark the area to be patched two to three inches outside the damaged area.

2.   Remove surface concrete with light- to medium-weight hammers to avoid damaging the sound concrete on the bottom layer of the pavement slab.

3.   Sandblast exposed concrete and clean the area with compressed air.

4.   For other than pre-cast, place a form for reestablishing the shoulder edge.

5.   In reinforced pavement (except for pre-cast repair) reestablish the reinforcement by overlapping and tying or welding with either a doubleface 4-inch weld or a single-face 8-inch weld.

6.   Brush in cement or epoxy grout.

7.   Place low-slump concrete with adequate mechanical vibratory screeds.

8.   Texture and cure the concrete.

9.   Apply a double application of white pigmented curing compound.

Temporary (asphalt) repair
1.   Blow out joints with compressed air.
2.   Remove broken concrete and square up the sides of the area.
3.   Apply a tack coat.
4.   Place an asphalt wedge and compact it.

Mud jacking:
      Mud jacking raises and adjusts a slab that has settled. Workable material is forced through holes drilled in the concrete slab. The material exerts sufficient pressure on the lower side of the slab to raise it before the material has traveled beyond the desired area:

1.   Examine the site and determine low spots by using line or elevation levels as appropriate. Look for the points of water intrusion into the pavement and where the subbase material has been deposited.

2.   Drill approximately 2-inch diameter core holes through the concrete slab at selected locations, or use preformed holes or previously drilled holes as appropriate.

3.   Starting at the downhill portion of the void and working up, begin pumping the mud jack mix into the predrilled holes. As the mixture raises the slab to the desired elevation or the void fills to capacity, move uphill to the next set of drill holes. It's important to lift the slab uniformly to avoid cracking it.

4.   Plug each hole temporarily once the hose is removed. Use a plastic plug or a burlap bag until the mixture has cured.

5.   After the entire slab area has been adjusted to grade, clean out each hole and refill with a fast-setting cement grout.

6.   Reseal any cracks and joints.

7.1.1.2—Asphalt Paving
Every 6 months, inspect and maintain asphalt paving as follows:

WHAT TO LOOK FOR	PROBABLE CAUSE	REMEDIAL ACTIONS
Rutting or surface depressions running parallel to traffic	If rut has raised edges, most likely cause is poor asphalt mix.	Full-depth repair or overlay.
	If rut has longitudinal cracks, most likely cause is failure of the sub-base.	
	Smooth ruts are typically caused by poor compaction and vehicle loading.	
	Progressive rutting (ruts that grow deeper and wide) typically results from very poor mix, poor sub-base, or inadequate road design.	

Fatigue cracking	Interconnected cracks (early) and loss of pavement (later) due to traffic loading.	Crack sealing, seal coat, or overlay.
Alligator cracking	Repeated traffic loads that cause high stress at the bottom of the pavement.	Crack sealing, seal coat, or overlay.
Washboards (ruts running transverse to road traffic)	Settlement of sub-base.	Full-depth repair or overlay.
Potholes	Water infiltrates through cracks and freezing, expanding upward against the pavement, which creates voids where vehicle loads cause pavement collapse. Later, vehicle loads will break away pavement at the edges, enlarging the pothole.	Surface patching or full-depth repair.
Pavement markings	Wear and deterioration	Apply new markings in accordance with Maintenance Procedure 7.1.1.4.

Crack sealing:

1. Saw-cut cracks and clean out with compressed air.

2. Apply low modulus rubberized asphalt sealant at 350-415°F.

3. Pour and even bead of sealant into the crack no higher than 1/2" above existing pavement surface.

4. To remove excess sealant, run a U-shaped squeegee or sealing shoe over the bead to flatten it.

5. Keep traffic off sealed cracks for 2-3 hours.

Surface patching

1. Before patching, correct drainage problems that caused the problem.

2. Clean the area to be patched with broom and compressed air. Remove any loose or deteriorated pavement.

3.   Fill the area to be patched with hot mix asphalt until it is slightly higher than surrounding existing pavement.

4.   Smooth the patch with a shovel or lute.

5.   Compact the patch using a hand or motorized tamper or vibratory roller.

6.   Keep traffic off patch until the asphalt is cool enough to touch.

Full-depth repair:
1.   Saw cut around the entire deteriorated area.

2.   Remove material manually or with a jack hammer. Once all material is removed, clean out all loose material with a shovel and then compressed air.

3.   Apply a tack coat to the edges of the existing pavement.

4.   Place new hot mix asphalt material in layers not exceeding 2.5" thick, smooth with a shovel or lute, and then compact with hand or motorized tamper (or, if large, with a vibratory roller).

5.   Do final compaction with vibratory roller.

6.   Keep traffic off patch until the asphalt is cool enough to touch.

Seal coat repair:
        Seal coat is an application of asphalt binder, followed by an application of aggregate. A good seal coat will have an adequate roadway crown; few unsealed cracks, potholes, or ruts; and a smooth, tight surface with no bleeding (binding coming to the surface) or raveling (a rough, pitted surfaced due to loss of aggregate). Seal coat surfaces as follows:

1.   Before seal coat, perform crack sealing and / or patching as necessary.

2.   Sweep and clean the pavement to remove debris.

3.   Spray the pavement with the binder. Use roofing paper to make a sharp line across the pavement when starting and stopping application. Align the binder application with the center line of the road. Align the nozzles and set the spray bar height as appropriate.

Apply only as much emulsion as the chip spreader will be able to cover with a load of aggregate. Calibrate this distance by measuring the distance the chip spreader travels on one load of aggregate. Apply binder at the rate of 0.25–0.3 gal/sy. If the existing pavement is smooth with few voids, the application rate is less. If the pavement is rough with lots of voids, the application rate is higher. The amount of applied should be approximately 50 to 70 percent of the thickness of the aggregate.

4. In general, apply the aggregate before the binder has set, usually within two to three minutes after it has been applied or before the surface has turned black. Apply aggregate at the rate of15–30 lb/sy. If the aggregate is not covering the binder, increase the aggregate application rate. If excess aggregate is visible, decrease the aggregate application rate.

5. Roll the aggregate with a pneumatic tire roller closely behind the chip (aggregate) spreader. Make two to four passes on a 24-foot wide roadway. Do not start and stop the roller quickly. This will cause diffraction of the surface.

6. The following day, sweep up excess unbound aggregate.

Overlay with hot mix asphalt:
1. Take special care to ensure that drainage will be maintained (e.g., bevel grind cross streets at flow lines).

2. Remove loose material and water from deteriorated areas. Clean, patch, and compact.

3. After the paving has been cleaned, apply a tack coat at the proper rate so as to avoid pushing or shoving the mat.

4. At the beginning of the overlay section, set up the paver to run the finish course. Set the heated screed on lath to gain prior mat elevation.

5. Back the truck up to the paver. When contact is made, raise the truck box before the tailgate is tripped to deliver the mix to the paver hopper. This dumping procedure will cause the mix to slide against the tailgate. Upon tripping the gate, the mix will flood the

hopper and reduce the amount of segregation that appears behind the screed.

6.   If the truck needs to be pulled away from the paver after loading (due to incline or some other reason), before the paver starts, make sure the mix has not spilled on the street and piled up in front of the paver tracks. Remove any spills. Otherwise the paver will ride on this material, and the pavement surface will be irregular.

7.   During paving, the hopper on the paving machine should be full at all times to ensure a constant flow of materials to the screed. In addition, the augers that move the mix in front of the screed should be turning most of the time so that the mix is uniform in density before compaction.

8.   Compaction is accomplished with a rubber-tired roller (breakdown roller) and a steel-wheeled roller (finish roller). The rubber-tired roller provides energy to compact the mix; it should be as close as possible behind the lay-down machine but not so close that the mix is rutted or disturbed. The finish roller removes the wheel tracks left by the rubber tired roller and should be as close as possible behind it without tearing the surface.

9.   After paving and rolling operations have concluded, barricade the street to allow for cure time overnight.

7.1.1.3—Curbs and Gutters
When inspecting paving, inspect and maintain curbs and gutter as follows:
1.   Replace curb and gutter section(s) that show settlement, uplift, vehicle damage, and/or general deterioration.

2.   Inspect drainage structures and repair as necessary. Pay careful attention to the mounting and support of cast iron grates to prevent them from becoming vehicle hazards.

7.1.1.4—Pavement Lines & Markings
When inspecting paving, inspect and maintain pavement lines and markings as follows:

1.   Temporary markings or temporary repairs to existing markings (providing 6-12 months life) may be made with traffic paint.

2.   For new markings or long term repair of existing markings, durable materials are required for longer life. Apply thermoplastics, which are applied as solids at ambient temperature and then melted, to provide the best life-cycle cost (5-7 year life).

7.1.1.5—Signage
Routinely, inspect and maintain signage as follows:
1.   Clean signs of graffiti, mud, etc. to ensure their visibility.

2.   If sign is damaged (shotgun pellets, graffiti that won't clean, accident, or vandalism), replace the sign.

3.   When replacing a sign, re-evaluate the current sign content, location, color, etc. and revise as needed to make sign more effective as a traffic control device. Make sure sign meet state standards. [All traffic control devices used on *public* streets and highways in must conform to the *Manual on Uniform Traffic Control Device*, Federal Highway Commission (2000), and local and state DOT requirements.]

7.1.1.6—Bridges
Annually, inspect each bridge and its approaches. If any of the following problems are identified, retain a Professional Engineer to provide a more detailed inspection and evaluation and use the engineer's report to adjust both the performance life of the bridge and its maintenance requirements.

Approaches:
On gravel road approaches, look for the following potential problems:
1.   Poor crown transition from the road to the bridge deck

2.   Too much aggregate and/or inadequate crust on the bridge approach, so that the aggregate migrates onto the bridge deck. Aggregate on the bridge deck may, in effect, narrow the operating width of the bridge.

3.   Standing water or erosion at the shoulder line.

On paved road approaches, look for the following potential problems:

1. Pavement distresses and excessive cracking.
2. Joint failures.
3. Erosion at the pavement edges.
4. Cracking or settlement of approach slabs.
5. Poor condition of expansion joint where the slab meets the bridge deck.
6. Poor "ride."

Deck:

On timber decks, look for the following potential problems

1. Loose nails, spikes, or fasteners
2. Openings between planks over abutments and piers which allow dirt to sift through.
3. Split, worn, broken, or decayed planks

On concrete decks, look for the following potential problems:

1. Cracking
2. Leaching
3. Exposed reinforcing
4. Scaling
5. Potholes
6. Spalling
7. Other evidence of deterioration

On steel decks, look for the following potential problems:

1. Corrosion
2. Unsound welds
3. Loose welds where the deck is fastened to the stringers
4. Dirt collected in open-grid decking on top of stringers
5. Deteriorated paint
6. Observe the condition of structural beams and trusses by sighting along the roadway rail or curb and along the truss chord members.

Look for truss misalignment, either vertical or horizontal. Bent trusses may reduce the bridge's operating width and/or reduce the structure's soundness. Note any members damaged by vehicles.

Underside of deck:

1.   Seepage, usually indicated by calcium deposits

2.   Cracks in the deck

3.   Exposed reinforcing

4.   Erosion at the bottom of the columns

5.   Deteriorated concrete in the columns

6.   Pier caps that are cracked or out of alignment

7.   Piers that are damaged due to ice or other debris

Slope protection:

On paved slope protection, look for the following potential problems:

1.   Broken panels (Broken panels may not need to be replaced if they are seated and generally conform to the slope.)

2.   Cracks that should be sealed to prevent water intrusion, which may cause settlement and/or sliding of the panels.

On riprap or revetment slope protection. look for the following problems:

1.   Bare areas

2.   Exposed fabric

3.   Erosion

4.   Inadequate rock size

Drainage systems:

Look for drainage system problems such as blockage, corrosion, erosion, etc.

Waterway:

Look for debris collecting near piers or in the stream channel. Debris accumulations may cause scour, redirect the stream channel, apply excessive hydraulic loads, or become a fire hazard.

7.1.4.0—Fencing & Gates

For high security applications, inspect fencing and gates weekly. For less secure applications, inspect fencing and gates monthly. For sports field fencing, inspect fencing and gates every 3 months. Maintain chain link and other metal fencing and gates as follows:

1. Washouts or other openings under the fence must be closed off by adding fencing or regrading to eliminate the opening.

2. Check and repair any fence posts that have deteriorated, rusted, or washed out.

3. Repair any openings, cuts, or sags in the fence fabric.

4. Replace missing or damaged barbed or razor wire on fence top guard (as applicable).

5. Inspect and repair gate components (hinges, tracks, swing clearances, latches and locks, etc.) as required.

6. For low ball field fences, inspect top safety padding and repair or replace as required.

7.1.4.1—Retaining Walls

Annually, inspect retaining walls and maintain as follows:

WHAT TO LOOK FOR	PROBABLE CAUSE	REMEDIAL ACTIONS
Wall cracking or bulging	Water pressure behind wall, foundation	Retain Professional Engineer to examine wall and recommend wall or mortar failure remedial action.
Clogged drains	Soil infiltration, plant growth.	If clogged with soil, route out with small auger or sharp metal probe. Remove plant growth that blocks drain.
Separation of top courses or cap	Mortar deterioration, freeze-thaw action, etc.	Repair.

7.1.4.2—Site Furnishings

Site furnishings include trash receptacles, benches, kiosks, tables, bicycle racks, playground equipment, and outdoor amenities that may be subject to weather deterioration, graffiti, and/or vandalism. Inspect site furnishings every two weeks and maintain as follows:

Tables:
1. Check that tables are clean, free of rust, mildew, and graffiti. Clean or repair as required.

2. Ensure that table hardware is intact.

3. Ensure that table frames are intact and slats are properly secured.

4. Ensure that table seats and tops are smooth with no protrusions and have no exposed sharp edges or pointed corners.

Trash Receptacles:
1. Check that receptacles are clean and are being routinely emptied.

2. Check that wood receptacles are painted and free of damage or missing parts.

3. Ensure that hardware for wood receptacles is intact.

4. Ensure that concrete receptacles are intact and free of cracks or damage. Repair as required.

5. Ensure that the area around trash receptacle is clean and free of trash and debris.

Wood Benches:
1. Determine if slats are smooth and structurally sound. Repair as required.
2. Ensure that hardware is intact and structurally sound.
3. Ensure that nails, bolts, or screws are flush with the surface.
4. Ensure that seats and backing are smooth with no protrusions and have no exposed sharp edges or pointed corners.

Metal Benches:

1. Ensure that hardware is intact and structurally sound.

2. Ensure that bolts or screws are flush with the surface.

3. Ensure that seats and backing are smooth with no protrusions and have no exposed sharp edges or pointed corners.

Drinking Fountains:

1. Ensure that fountains are accessible and operational. Test and repair any operating malfunction.

2. Ensure that fountains are installed on solid surfaces and free of standing water and debris.

7.1.4.3—Water Features

Inspect and maintain fountains, shallow pools, ponds, and other water features as follows:

1. Twice a week, check pool for algae growth. If algae growth is a problem, and the water feature does not incorporate plants and fish or other wildlife [feature is decorative only], chemical water treatment can be used. To protect plants and animal life, an aerator can be used in lieu of chemicals to increase oxygen levels. [An aerator can be constructed from a submersible pump with its discharge below water. Run the pump at night or on weekends as needed to reduce algae. For a large basin, it may be necessary to have more than one aerator.]

2. Weekly, clean pump inlet screen and check pump for proper operation. Check water level so that pump inlet remains submerged.

3. Annually, service pump. If submersible type, remove pump. Remove pump housing and remove and clean impeller. [If needed, a 50/50 solution of "C.L.R." and water can be used. Soak impeller for 20 minutes.]

4. Annually, clean fountain nozzles (as applicable) in accordance with manufacturer's written instructions.

7.1.4.4—Flagpoles

Inspect and maintain flagpoles as follows:

1.   Inspect poles annually and service as required:
     a. Aluminum or stainless steel poles require only cleaning based upon inspection.
     b. Re-paint steel poles every 3-7 years based upon inspection.
     c. Check condition of anchor bolts, nuts, and washers. Clean and liberally spray with outboard engine anti-corrosion coating.

2.   Inspect halyards (ropes) quarterly and replace as needed with braided polypropylene or nylon.

3.   Inspect fittings quarterly. Replace as indicated by inspection.

7.1.4.5—Playgrounds/Equipment
Maintain playgrounds and playground equipment as follows:

Daily:
Inspect for and remove debris or litter such as tree branches, cans, bottles, broken glass, etc. from the play area.

Monthly:
1.   Inspect and make sure that protective surfaces around playground equipment are adequate and not deteriorated:
     a. At least 12" thick mulch, sand, or pea gravel. Ensure mulch has not degraded.
     b. Surface must extend at least 6 feet in all directions around equipment. For swings, surface must extend twice the bar height, front and back.
     c. Any surface that is inadequate or has deteriorated must be enhanced or replaced.
     d. If sand is the protective covering, treat with a solution of 15 oz of sodium hypochlorite to 5 gallons of water for every 300 cubic feet of sand. Apply with a watering spray and then rake in.

2.   Inspect equipment for condition and safety hazards and repair as required, as follows:
     a. Sharp points, corners, or edges on the equipment.
     b. Missing or damaged protective caps or plugs.
     c. Hazardous protrusions or projections.
     d. Potential clothing entanglement hazards such as open S-hooks

or protruding bolts.

e.   Pinch, crush, or shearing points or exposed moving parts.

f.   Trip hazards such as exposed footings or anchoring devices, rocks, roots, other obstacles.

g.   Rust, rot, cracks, or splinters, especially where equipment is in contact with the ground.

h.   Deteriorated, broken, or missing components on equipment, such as handrails, guardrails, protective barriers, steps or rungs on ladders, etc.

i.   Equipment anchoring.

7.1.5.0—Erosion/Runoff Control

Inspect and maintain erosion or runoff control features as follows:

Irrigated Grass Buffer Strip:

Required Action	Maintenance Objective	Frequency of Action
Lawn mowing	Maintain a dense grass cover at a recommended length of 2 to 4 inches. Collect and dispose of cuttings offsite or use a mulching mower.	As needed or recommended by inspection.
Lawn care	Use the minimum amount of biodegradable, nontoxic fertilizers and herbicides needed to maintain dense grass cover, free of weeds. Reseed and patch damaged areas.	As needed.
Irrigation	Adjust the timing sequence and water cover to maintain the required minimum soil moisture for dense grass growth. Do not overwater.	As needed.
Litter removal	Remove litter and debris to prevent gully development, enhance aesthetics, and prevent floatables from being washed offsite.	As needed by inspection.
Inspections	Inspect irrigation, turf grass density, flow distribution, gully development, and traces of pedestrian or vehicular traffic and request repairs as needed.	Annually and after each major storm (greater than 0.75" precipitation).

Turf replacement	To lower the turf below the surface of the adjacent pavement, use a level flow spreader, so that sheet flow is not blocked and will not cause water to back up onto the upstream pavement.	As needed when water padding becomes too high or too frequent a problem. The need for turf replacement will be higher if the pavement is sanded in winter to improve tire traction on ice.
Inspections	Check the grass for uniformity of cover, sediment accumulation in the swale, and near culverts.	Annual inspection.

Grass-Lined Swale:

Required Action	Maintenance Objective	Frequency of Action
Lawn mowing and Lawn care	Maintain irrigated grass at 2 to 4 inches tall and nonirrigated native grass at 6 to 8 inches tall. Collect cuttings and dispose of them offsite or use a mulching mower.	As needed.
Debris and Litter removal	Keep the area clean for aesthetic reasons, which also reduces floatables being flushed downstream.	As needed by inspection, but no less than two times per year.
Sediment removal	Remove accumulated sediment near culverts and in channels to maintain flow capacity. Replace the grass areas damaged in the process.	As needed by inspection. Estimate the need to remove sediment from 3 to 10 percent of total length per year, as determined by annual inspection.
Grass reseeding and mulching	Maintain a healthy dense grass in channel and side slope.	As needed by annual inspection.
Inspections	Check the grass for uniformity of cover, sediment accumulation in the swale, and near culverts.	Annual inspection.

Porous Landscape Detention:

Porous Landscape Detention

Required Action	Maintenance Objectives	Frequency
Lawn mowing and vegetative care	Occasional mowing of grasses and weed removal to limit unwanted vegetation. Maintain irrigated turf grass as 2 to 4 inches tall and nonirrigated native turf grasses at 4 to 6 inches.	Depending on aesthetic requirements.
Debris and litter removal	Remove debris and litter from detention area to minimize clogging of the sand media.	Depending on aesthetic requirements.
Landscaping removal and replacement	The sandy loam turf and landscaping layer will clog with time. This layer will need to be removed and replaced, along with all turf and other vegetation growing on the surface, to rehabilitate infiltration rates.	Every 5 to 10 years. May need to do it more frequently if exfiltration rates are too lowl.
Inspections	Inspect detention area to determine if the sand media is allowing acceptable infiltration.	Bi-annual inspection of hydraulic performance.

Extended Detention Basin:

Required Action	Maintenance Objective	Frequency of Action
Lawn mowing and lawn care	Occasional mowing to limit unwanted vegetation. Maintain irrigated turf grass as 2 to 4 inches tall and non-irrigated native turf grasses at 4 to 6 inches.	Depending on aesthetic requirements.
Debris and litter removal	Remove debris and litter from the entire pond to minimize outlet clogging and improve aesthetics.	Including just before annual storm seasons and following significant rainfall events.

Erosion and sediment control	Repair and revegetate eroded areas in the basin and channels.	Periodic and repair as necessary based on inspection.
Structural	Repair pond inlets, outlets, forebays, low flow channel liners, and energy dissipators whenever damage is discovered.	Repair as needed based on regular inspections.
Inspections	Inspect basins to insure that the basin continues to function as initially intended. Examine the outlet for clogging, erosion, slumping, excessive sedimentation levels, overgrowth, embankment and spillway integrity, and damage to any structural element.	Routine – Annual inspection of hydraulic and structural facilities. Also check for obvious problems during routine maintenance visits, especially for plugging of outlets.
Nuisance control	Address odor, insects, and overgrowth issues associated with stagnant or standing water in the bottom zone.	Handle as necessary per inspection or local complaints.
Sediment removal	Remove accumulated sediment from the forebay, micro-pool, and the bottom of the basin.	Every 10 to 20 years, as necessary per inspection…more often as needed. The forebay and the micro-pool will require more frequent cleanout than other areas of the basin, say every 1 or 2 years.

Sand Filter Detention Basin:

Required Action	Maintenance Objectives	Frequency
Debris and litter removal	Remove debris and litter from detention area to minimize clogging of the sand media.	Depending on aesthetic requirements.
Landscaping removal and replacement	If the sand filter is covered with rock mulch, bluegrass, or other landscaping covers, the cover must be removed to allow access to the sand media. Replace	Every 2 to 5 years.

	landscaping cover after maintenance of sand media is complete.	
Scarify filter surface	Scarify top 3 to 5 inches by raking the filter's surface.	Once per year or when needed to promote drainage.
Sand filter removal	Remove the top 3 inches of sand from the sand filter. After a third removal, backfill with 9 inches of new sand to return the sand depth to 18 inches. Minimum sand depth is 12 inches.	If no construction activities take place in the tributary watershed, every 2 to 5 years depending on observed drain times, namely when it takes more than 24 hours to empty 3-foot deep pool. Otherwise more often. Expect to clean out forebay every 1 to 5 years.
Inspections	Inspect detention area to determine if the sand media is allowing acceptable infiltration.	Bi-annual inspection of hydraulic performance, one after a significant rainfall.

Constructed Wetlands Basin:

Required Action	Maintenance Objective	Frequency of Action
Lawn mowing and lawn care	Mow occasionally to limit unwanted vegetation. Maintain irrigated turf grass at 2 to 4 inches tall and nonirrigated native turf grasses at 4 to 6 inches.	Depending on aesthetic requirements.
Debris and litter removal	Remove debris and litter from the channel.	Including just before annual storm seasons (that is, in April and May) and following significant rainfall events.
Sediment removal	Remove accumulated sediment and muck along with wetland vegetation growing on top of it. Re-establish growth zone depths and revegetate with original wetland species.	Every 10 to 20 years as needed by inspection if no construction activities take place in the tributary watershed. More often if they do.

Aquatic plant harvesting	Cut and remove plants growing in wetland (such as cattails and reeds) to remove nutrients permanently with manual work or specialized machinery.	Perform this task once every 5 years or less frequently as needed to clean the wetland zone out.
Inspections	Observe inlet and outlet works for operability. Verify the structural integrity of all structural elements, slopes, and embankments.	At least once a year, preferably once during rainfall event of more than 1.5", resulting in runoff.

Retention Pond Basin:

Required Action	Maintenance Objective	Frequency of Action
Lawn mowing and lawn care	Mow occasionally to limit unwanted vegetation. Maintain irrigated turf grass 2 to 4 inches tall and non-irrigated native turf grasses at 4 to 6 inches.	Depending on aesthetic requirements.
Debris and litter removal	Remove debris and litter from the entire pond to minimize outlet clogging and aesthetics. Include the removal of floatable material from the pond's surface.	Including just before annual storm seasons (that is, April and May) and following significant rainfall events.
Erosion and sediment control	Regrade and revegetate eroded and slumped areas above the pond and along channels. Repair damaged inlet and outlet energy dissipators.	Periodic and repair as necessary based on inspection.
Inspections	Inspect the retention pond for functioning as initially intended. Pay attention to outlet clogging. Also note erosion, slumping, sedimentation levels, overgrowth, embankment and spillway integrity, and damage to structural elements of the facility.	Annual inspection of hydraulic and structural facilities. Biannual performance and maintenance inspections.

Nuisance control	Address odor issues, insects, and overgrowth with appropriate measures.	As necessary per inspection or local complaints.
Structural repairs	Repair such items as inlet/outlet works and energy dissipator liners. Stabilize banks and berms. Repair damage caused by larger storm events.	As necessary per inspection.
Sediment removal	Empty the pond, divert the base flow, and dry out bottom sediments in fall and winter months to allow access with backhoe. Remove accumulated sediment along with aquatic growth overlaying them. Re-establish original design grades and volumes and replant aquatic vegetation.	As indicated per inspections and sediment accumulation. Expect to do this every 10 to 20 years if no construction activities take place in the tributary watershed. More often if they do. Expect to clean out the forebay every 1 to 5 years.
Aquatic growth harvesting	Remove aquatic plants such as cattails or reeds, which also permanently removes nutrients. Use an aquatic harvester and dispose of the material offsite.	Perform every 5 to 15 years or as needed to control their accumulation.

Constructed Wetlands Channels:

Required Action	Maintenance Objective	Frequency of Action
Lawn mowing and lawn care	Mow occasionally to limit unwanted vegetation. Maintain irrigated turf grass at 2 to 4 inches tall and nonirrigated native turf grasses at 4 to 6 inches.	Depending on aesthetic requirements.
Debris and litter removal	Remove debris and litter from the channel.	Including just before annual storm seasons (that is, in April and May) and following significant rainfall events.

Sediment removal	Remove accumulated sediment and muck along with wetland vegetation growing on top of it. Reestablish growth zone depths and revegetate with original wetland species.	Every 10 to 20 years as needed by inspection if no construction activities take place in the tributary watershed. More often if they do.
Aquatic plant harvesting	Cut and remove plants growing in wetland (such as cattails and reeds) to remove nutrients permanently with manual work or specialized machinery.	Perform this task once every 5 years or less frequently as needed to clean the wetland zone out.
Inspections	Observe inlet and outlet works for operability. Verify the structural integrity of all structural elements, slopes, and embankments.	At least once a year, preferably once during rainfall event of more than 1.5", resulting in runoff.

7.1.5.1—Irrigation Systems

Irrigation systems should be operated and maintained as follows:

1. To the maximum extent possible, underground low volume irrigation systems, such as micro-sprayers or drip tubes, shall be utilized to minimize evapotranspiration losses. All planting beds, shrubbery, etc. can be efficiently irrigated this way.

2. Sprinkler systems should be limited to turfgrass irrigation only, should be designed to achieve at least 80% planting coverage, minimizing sprinklering impervious surfaces and the associated runoff losses. Coverage can be tested by placing empty cans or cups in a uniform 10-15 foot square pattern in the sprinklered area. By comparing the water level in each can or cup, flows and patterns can be adjusted. Sprinkler systems must be controlled on the basis of soil moisture sensors, evapotranspiration (ET) controllers, etc. Wind speed sensors should be used to shut-off water flow when wind speeds exceed 10 mph. *Time-clock controls should not be utilized as the sole method of irrigation system control.*

Monthly during the irrigation season, inspect and maintain irrigation systems as follows:

Controller:

1. Open the cabinet for the irrigation controller and make sure it is free of debris such as cobwebs or dirt. Replace the battery.

2. Check all wiring connections for wear and breakage. Repair if necessary.

3. Check controller sensors and controller for proper operation. Test sensor calibration.

Sprinkler System:

1. Flush system: Before running the system, remove the last sprinkler head in each line and let the water run for a few minutes to flush out any dirt and debris. Replace the sprinkler heads and turn the system on, running one valve at a time.

2. Look for obviously broken or clogged heads and make the necessary repairs. Consider installing irrigation heads that have screens to prevent debris (grass, soil, or bugs) from clogging the sprinkler heads. Clean out screens that may be clogged.

3. Observe the lowest head in each station for leaks. Algae or moss may be growing in the area and may indicate the problem.

4. Look for a very fine mist from spray heads, which is caused by excessive pressure in the system. Correct the problem with a pressure regulator downstream of the water meter, pressure regulating sprinkler heads, or added throttling valves on individual sprinkler heads.

5. Check to see if the sprinklers are covering the desired area uniformly. If your pressure is too low, modify system so there are fewer sprinklers on each valve.

6. Check to see that irrigated areas have at least 80% coverage and there is no overspray onto impervious surfaces. Adjusting the spray pattern or replace the spray nozzle(s) with another that has the correct spray pattern.

7. Check to see if water is draining through the lower heads. Install check valves where appropriate, or replace existing heads with heads that contain built-in check valves.

8. Check to see that different types of heads are not used in the same irrigation zone. Nozzles should also be correlated for matched precipitation rates.

9. Look for over-spray of sprinklers onto sidewalks, driveways, and streets. The sprinklers' spray patterns should either be adjusted or changed to a pattern that will stay within the planting area.

10. Look for blocked spray patterns. Remove vegetation and other obstructions that may be blocking the spray, or consider raising the heads.

11. Check each head to see that it is at ground level. Raise sunken heads to grade or replace existing short pop-up heads in the lawn with taller pop-ups, as necessary. You can also trim around existing heads to avoid blocking the spray but you will have to do this on a continual basis.

12. Heads should be aligned vertically, except in sloped areas. In a sloped area, heads should be aligned perpendicular to the slope to achieve proper coverage. Tilted heads can cause ponding and uneven coverage.

13. Check to see if you have head to head coverage between sprinklers. If necessary, consult a qualified professional to design a system with head-to-head spacing.

14. Check for piping leaks, usually indicated by water welling up in an area or little or no flow to specific heads. To find the leak, turn off the water supply at the highest elevation and start digging down hill. Repair the leak.

Drip System:
1. Clogged emitters should be replaced. If the system does not have a water filter, one should to be installed.

2. Check the placement of emitters. Emitters need to be at the edge of the root-ball on new plantings and moved to the drip line (edge of foliage) of established plants.

3. Check for emitters that have popped off tubing because of high pressure. Install a pressure regulator on the valve for all drip stations.

4. Check to see that all of your emitters are in place. Missing and broken emitters need to be replaced to keep your system running efficiently.

5. Look for pinched or broken tubing and straighten or replace it.

6. Make sure all tubing is attached to the appropriate emitters and that connections are secure.

Winterization:
At the end of the irrigation season, shut off irrigation system and prepare for winter as follows:

1. Shut off water supply and switch irrigation controller to "off."

2. Open each zone or area drain valve or, if necessary, remove a head at the end of each branch to drain piping. If system does not drain adequately, use low pressure compressed air (50 psig max) to force water from the piping. [Close valves at BFD to avoid damage due to excess pressure.]

SITE CIVIL AND MECHANICAL UTILITIES

7.2.1.0—Well Water
Test well water every 2 years for bacteria, heavy metals, and volatile organic compounds. If the well is located near industrial facilities, airports, service stations, dry cleaners, auto shops or dealers, large retail facilities, or other "point" sources for groundwater contamination, test annually. *If the water tests positive for any of these, cease use of the well and investigate cause of contamination.*

7.2.1.1—Well System
Annually inspect well water system including above ground pump, pressure tank and controls, water softener, and piping and valves.

Submersible pump:
1.   Check and tighten pump electrical connections at the wellhead and at the disconnect.

2.   Test motor amps when pump is on and compare to nameplate data.

3.   Evaluate water flow rate by opening several faucets and watching pressure condition at tank. When pressure drops to low setting and pump starts, close faucets and measure the amount of time it takes for the pump to cycle off. If this exceeds 30 seconds, the pump flow may be inadequate.

4.   Pull and examine submersible pump every 5 years. Replace pump every 15 years or as indicated by inspection.

Above ground pump:
1.   Check and tighten pump electrical connections at the pump and at the disconnect.

2.   Test motor amps when pump is on and compare to nameplate data.

3.   Evaluate water flow rate by opening several faucets and watching pressure condition at tank. When pressure drops to low setting and pump starts, close faucets and measure the amount of time it takes for the pump to cycle off. If this exceeds 30 seconds, the pump flow may be inadequate.

4.   Replace pump every 15 years or as indicated by inspection.

Pressure tank and controls:
1.   Check water tank for corrosion, leaks, etc. that require repair.

2.   Check water pressure gauge accuracy with a test gauge and compare pump controller settings (on and off) against gauge reading. Replace gauge and/or adjust controller settings as required. If controller malfunctions, replace.

3. Test to determine if the tank is water-logged by tapping tank from bottom to top with the handle of a screwdriver. If the sound indicates that there is no air cushion, follow tank manufacturer's written instructions for draining, refilling, and pressurizing tank.

Water softener:

1. Fill with sodium chloride to keep reservoir at least 1/3 full at all times [to maintain both good softening and prevent "salt bridging."] For locations where occupants must limit sodium intake, use potassium chloride in lieu of sodium chloride.

2. Clean the injector:
    a. Turn the water softener control to the bypass mode (so the main water line does not run through the softener). On some units, you also may need to disconnect the unit from the electrical source.
    b. Unscrew the softener injector cap and remove the screen and injector nozzle. You may need the owner's manual to determine where these components are located on your model.
    c. Remove and clean the screen with warm soapy water, then rinse in clear water. If it is damaged, replace the screen.
    d. Clean the nozzle with canned air, a small wire, or a paper clip.
    e. Reassemble and test the unit.

3. Flush the brine line:
    a. Turn the water softener control to the bypass mode. On some units, you also may need to disconnect the unit from the electrical source.
    b. Loosen the brine line fittings at the injector housing and salt tank with a wrench, then remove the brine line.
    c. To clear a clog, use a large syringe or a turkey baster to inject warm water into the line.

4. *Do not use iron-removing additives unless specifically approved by the softener manufacturer.* A water softener WILL NOT remove any iron that makes the water cloudy or rusty as it comes from the faucet (called "red water iron"). To take red water iron out of water, an iron filter or other equipment is needed. If water supply has "clear water iron," clean resin bed every 6 months, or more often if iron appears

in the soft water between treatments. [Follow manufacturer's written instructions to clean the resin bed.]

Well head, piping, and valves:
1.   Annually, check well cap, well house or cover, insulation, heater (as applicable) for condition and operation. [The vent should have a screen over the vent hole to prevent insects and rodents from entering the well. In most cases a vent is needed to help a well produce water more efficiently, but can sometimes be plugged in lower-use domestic wells with little noticeable affects. The best type of vents are the ones which allow a little air to enter from the bottom of a U tube, thus preventing things spilled, dumped, or dropped onto the vent from entering the well.] Check all exposed piping and valves for leaks, cracks, etc. Repair or replace components as indicated by inspection.

2.   Inspect the ground around the casing to check for slumping and settlement. Backfill slumped holes around the well casing with compacted clay soil. The well head should be at least 8" above grade and the land surface around the well casing should slope away from the well to prevent the ponding of surface water.

3.   Make sure that things are not kept around the well that could release contaminants to the well [such as fuel cans, fertilizer, pesticide containers, paint, dog or animal pens, gasoline- and diesel-powered tools and vehicles, and solvents.]

4.   Pull and examine foot valve every 5 years. Examine suction lines and return lines for proper connections, splits or cracks, or brittleness. Repair or replace components as indicated by inspection.

7.2.2.0—Sanitary Sewer System
Inspect, clean, and maintain sanitary sewer system as follows:

Spill response:
1.   Respond immediately to minimize the amount of sewage entering the storm water drainage system. Block storm drain inlets and/or contain sewage by spill control barriers.

2.   Remove the sewage using vacuum equipment.

3.   Perform field tests to determine the source of the leak or spill and perform repairs.

4.   If leaks or spills occur regularly on a specific line, install liner or replace piping.

Sewer line cleaning:
1.   Annually, inspect manholes for debris build-up and condition. Clean out as needed. Check for cracks or other structural problems and repair as required.

2.   Every 5 years (or more frequently for kitchen discharge lines), clean sewer lines to remove grease, grit, and other debris.

7.2.2.1—Septic System
Annually, inspect and maintain septic system as follows:
1.   Locate tank and manhole covers. [If tank is not fitted with manhole risers to bring access to grade level, install them.]

2.   Open manhole cover and inspect as follows:
     a.   Scum should be firm, with a crust, but not solid. It should be like pudding, a medium brown color, and 3-4" thick. Test scum with a stick or rod to determine thickness. If scum is too thick or does not appear to have the correct consistency, arrange to have tank pumped.
     b.   Test depth of sludge by slowly inserting a concrete hoe until resistance is felt. If sludge depth is at or above 50" of tank depth, arrange to have tank pumped.
     c.   If tank must be pumped, ensure that tank is pumped clean. Use a pressure washer if necessary to remove all sludge.
     d.   Check inlet and outlet tees to make sure they not broken or clogged.

3.   Visually inspect drainage field area. It plant growth is relatively lush and there are no sunken or overly wet areas, the field is probably functioning properly. If there is any doubt, have drain lines tested for clogs or other blockages with a water test.

7.2.2.2—Sewage Lift Station
Inspect and maintain sewage lift station as follows:

1. Weekly, inspect and test light and alarm systems. Visually inspect control panel wiring for obvious signs of electrical problems such as burned wiring, loose connections, or burn spots on cabinet. Repair as required.

2. Monthly, take amp readings on each pump motor. If the amp readings do not meet the manufacturer's specifications, it is an indication that debris is lodged in the propeller within the motor, or that water has entered the motor housing or the wiring.

3. Monthly, check generator (as applicable):
    a. Check oil, water, and fuel levels.
    b. Inspect hoses and belts.
    c. Check battery level and operation of charging system.
    e. Check that generator is warm.
    f. Every 3 months, test generator under pump load.

4. Every 3 months, clean any grease build-up on floats and test them to ensure proper performance.

5. Every 3 months, or when runtime hours of duplex pumps are not within 10% of each other, inspect submersible pumps to ensure that the impeller is free of debris. [If missing, install runtime hour meters on each motor will give one an accurate record of how often each motor is cycling and, thus, the amount of water being pumped through the system.] Run pumps in manual mode and listen for unusual noise.

6. Every 6 months, or more often if necessary:
    a. Pump out and clean wet well to prevent solids and grease build-up.
    b. Inspect check valves ensure proper working order and to prevent backflow from the force main to the wet well.

    c. Inspect all electrical motor control equipment to find poor connections and worn parts. Tighten connects and make repairs as indicated by inspection.

7.2.2.3—Waste Treatment System

Inspect and maintain package waste treatment system in accordance with manufacturer's written instructions and regulations of the local and state health and/or environmental departments.

7.2.2.4—Storm Water System

Inspect and maintain storm water system as follows:

1. Annually, and after every rain event exceeding 0.75″ of precipitation, inspect catch basins, headwalls, and manholes for debris build-up and condition. Clean out as needed. Check for cracks or other structural problems and repair as required.

2. Every 5 years (or more frequently if routine back-up occurs), clean drain lines to remove debris.

SITE ELECTRICAL UTILITIES

7.3.1.0—Exterior Lighting

Inspect exterior lighting (fixtures, lamps, sensors, poles, and stanchions) every 6 months.

1. Check lighting controller for proper settings and operation. Set decorative lighting [such as façade lighting, walkway lights, etc.] time clock controls to turn on at approximately 30 minutes before sunset and turn off at midnight. Security lighting should operate on basis of photocell, with no time clock control.

2. Clear away tree branches, bushes, etc. that block lighting coverage.

3. Check condition of poles or stanchions, including foundations. Repair or replace as indicated by inspection. Check wiring connections and tighten.

4. Every 20,000 hours of use [5-6 years], relamp all exterior fixtures. During relamping, clean both light fixture and lens. Replace photocell(s) during relamping with solid state type devices that do not loss sensitivity over time [avoid use of cadmium sulfide cells].

Appendix H

Preventative Maintenance For Access by Disabled Persons

Based on legal decisions relative the application of the requirements of the *Americans with Disabilities Act* (ADA), building owners are required to maintain all access features for persons with mobility and/or vision disabilities in fully operational condition, including but not limited to the following:

1. Maintain exterior pathways and repair any surface irregularities that become greater than ½ inch due to wear or cracking, and shall make other repairs to keep pathways from causing hazardous conditions. Inspect site each month and correct any deficiencies within 14 days of their discovery.

2. Maintain disabled parking spaces to have all appropriate signage and to keep access aisles to the spaces and to the main entrances they serve clear and usable. Inspect site each month and correct any deficiencies within 21 days of their discovery.

3. Maintain and replace as required all building signage that would direct persons with disabilities to the accessible paths of travel. Inspect facilities each month and shall correct any deficiencies within 21 days of their discovery.

4. Maintain all doors providing primary accessibility to be fully operable and unlocked during normal hours of operation of the facility and during all public functions whenever the primary entrance to the facility is unlocked. Maintain all access features such as accessible restrooms, elevators and platform lifts so that they are

fully operable and unlocked during normal hours of operation of the facility, and during all public functions. For any elevators or restrooms that are locked, provide keys to any such locked access features to disabled persons who request keys.

5.   Maintain all accessible plumbing fixtures, including toilets, urinals, lavatories, sinks, faucets, showers, and drinking fountains, to be fully operational and in compliance with ADA Accessibility Guide (ADAAG). Inspect facilities each month and repair any non-operating conditions within 24 hours of their discovery.

6.   Maintain all toilet accessories to be fully operational. Maintain all grab bars to be tight and structurally sound. Such features to be checked monthly and to be repaired within 24 hours of the discovery of any problem.

7.   Automatic door openers shall be checked every morning and fixed within 24 hours of the discovery of a problem.

8.   All safety or warning striping shall be inspected annually and repainted, repaired, or replaced as necessary.

Index

Printed in the United States
by Baker & Taylor Publisher Services